THE LÉGION DU NORD
1806-1808
Memoir of Major Coqueugniot

By
Major Lazare Claude Coqueugniot
Major of the Legion

Translated and Annotated
by G.F.Nafziger

Original work:
Paris

Aux Bureaux de la Nouvelle Revue Rétrospective
55, Rue de Rivoli, 55

. . . .
1898

This edition:

The Legion Du Nord: 1806 – 1808 by Major Lazare Claude Coqueugnoit
Translated by G. F. Nafziger
Cover by Richard Knotel - Legion of the North Fusilier, 1807
This edition published in 2022

The Nafziger Collection is an imprint of

Winged Hussar Publishing, LLC
1525 Hulse Rd, Unit 1
Point Pleasant, NJ 08742

Copyright © Winged Hussar Publishing
ISBN 978-1-950423-98-9 HC
ISBN 978-1-958872-04-8 Epub
LCN 2022947682

Bibliographical References and Index
1. History. 2. Napoleon. 3. Poland

Winged Hussar Publishing, LLC All rights reserved
For more information
visit us at www.whpsupplyroom.com

Twitter: WingHusPubLLC
Facebook: Winged Hussar Publishing LLC

The Legion du Nord

Notes on this Edition and its Translation:

The original work is titled *La Légion du Nord, 1806-1808, Mémoire de Lazare Claude Coqueugniot, Major de la Légion*. It was published, earlier, in the 1898 edition of the magazine *Nouvelle Revue* retrospective.

The original footnotes have been included, without comment. Where additional information was necessary, or an explanation required, it was inserted, as a footnote, with the word, "Editor:" beginning the footnote.

The style of the writing was rather clipped, and I have taken the liberty of smoothing it out. When words were added to the text that were best included in the flow of the work than included as a footnote, they have been inserted in square brackets.

Titles that are easily translated into English were generally translated, but unique French terms, i.e., chef de bataillon (battalion chief), were not translated.

———

The Légion du Nord (1806-1808)
A Memoir of Lazare-Claude Coqueugniot[1]

---

[1]Communication from M. Gabreil Cottereau. – The manuscript is entitled, *History of the Legion of the North*, recounted in 1809 and prepared in 1819 by L.C. Chevalier Coqueugniot, who recruited and organized this corps in 1806; who had commanded it before the enemy until the Treaty of Tilsit, and who did not leave it until 1808.

Born on 30 October 1760 at Allerey (Côte d'Or), Lazare-Claude Coqueugniot enlisted as a simple soldier in the Piémont Infantry Regiment in 1780 and became a sous-lieutenant in 1791. He became a lieutenant in 1792, a captain in 1793, chef de bataillon in 1794, major (lieutenant colonel) of the 1st Légion du Nord in 1806, passed with this corps into the service of the Grand Duchy of Warsaw in 1807, colonel on 1 January 1808, reentered French service as an assistant inspector supernumerary of reviews in 1808, and was attached to the Minister of War in 1810 and in 1811. He was charged with the administration of the troops of the Grand Duchy of Berg in 1812 and was employed at Mézières in 1814. He retired in 1815 and died in 1829. Coqueugniot participated in the campaigns of 1782 and 1783 at sea, those of 1792 to 1801 in various armies, those of the Grande Armée in 1806 and 1807, that of Poland in 1808, and that of Germany in 1809. He was a member of the Legion of Honor from 17 September 1807 and Knight of the Empire from 31 January 1810. A decree of 8 September 1808 accorded him a grant of 2,000 francs rent on the Department of Trasimène.

He published in 1822 the *Mémoire historique sur les anciens monuments militiares de la ville de Strasbourg*, which lacked neither erudition, nor originality. He equally left behind a manuscript entitled *Histoire de l'armée et de son administration*, which belongs to M. Cottereau.

Hostilities against Prussia began on 9 October 1806 and, the following day, on 10 October, the Emperor named me major of the 1re Légion du Nord, whose cadre was to be organized in Landau.

I was not, at first, very happy with this commission, because I feared finding myself under the orders of an ignorant colonel, which would prevent me from putting the few talents and experience that I had gained in the infantry service into practice; however, two colonels were summoned, from Paris, who were Polish généraux de division in the service of France; that is, General Zajączek [2], who had served in Egypt for the 1st Legion and General Henry Wolodkowicz[3] for the second, whose cadre was being organized in Nürnberg.

This filled me with pleasure, because I knew that General Zajączek was a man of letters and who, consequently, would do me the justice I merited.

I left Martigues as soon as I received my commission and I arrived promptly at Strasbourg, where my family had always had its home, since I had married. There I found M. Lebrun, quartermaster-treasurer of the Legion. He informed me that the cadre of this corps, consisting only of the 1st Battalion, which was far from complete, had left Landau to move to Mainz, where there were currently a few thousand Prussian prisoners of war. I learned from M. Lebrun that General Zajączek waited for me with the greatest impatience, because it was necessary to recruit [the manpower for the Legion] from among the prisoners of war. The quartermaster had come to Strasbourg only to end the departure review with the inspector and to receive funds.

Adjutant Général Hennin de Cuvilliers had been provisionally placed in the Legion, as they awaited my arrival, to second General Zajączek, but he did not know

---

[2] Józef Zajączek, général de brigade on 11 May 1801 and général de division on 16 May 1812.
[3] Henry Wolodkowicz, général de brigade on 23 September 1806.

	Designation of Grades / Battalions		Officers	NCO's /Soldiers
Staff	General de division		1	-
	Commander		1	-
	Major 2nd in Command		4	-
	Chefs de bataillon		4	-
	Adjudant majors		1	-
	Quartermaster Treasurer		1	-
	Surgeon Major		4	-
	Surgeon aide Major		4	-
	Surgeon under Aide Majors		-	4
	Adjutant NCOs		-	1
	Drum Major		-	2
	Drum Corporal		-	8
	Musicians		-	3
	Artisans		-	1
	TOTAL		**20**	**19**
Company	Captain		1	-
	Lieutenant		1	-
	Sous-Lieutenant		1	-
	Sergeant Major		-	1
	Sergeant		-	4
	Corporals		-	8
	Quartermaster Corporal		-	1
	Drummer		-	2
	Fusiliers		-	120
1st Battalion	1st Carabiner Co	139	27	1,124
	7 Chasseur Co's	973		
	1 Voltiguer Co	139		
2nd Battalion	Ditto		27	1,124
3rd Battalion	Ditto		27	1,124
	TOTAL		**128**	**4,914**
	TOTAL			**5,043**

enough about the organization of a troop nor of the various services to be of great use.

The quartermaster gave me the ministerial table, indicating the composition of the Legion. It was as follows:

When the battalion arrived in Mainz the prisoners were divided into groups that were then designated as companies. The Emperor had assigned, to this place, the old Maréchal Kellermann, to whom detachments were sent that were to join the Grande Armée to replace the losses it had suffered. The maréchal only forwarded, with all promptness, the troops that arrived at Mainz, whether or not they were fully equipped and armed. This is why he did not give General Zajączek the time to establish the nominative lists of the portions of prisoners who would represent the companies of the Legion.

However, he wanted to lend to this general ten officers of the National Guard of the upper Rhine, which found themselves in Mainz at that time. These officers, added to those already in the cadre of the first battalion, only provided the number to form two companies and this was even more ridiculous, as there were still no sergeants or corporals for these troops. (It would have been unbecoming to give these ranks to the soldiers, who were prisoners of war and who did not know how to read or write, so General Zajączek asked the Minister War to take, from French regiments, soldiers and corporals capable of being employed as non-commissioned officers.

However, as Maréchal Kellermann required that the troops cross the Rhine immediately, they entered Germany having only one officer for two companies, plus Chef de Bataillon Vanrosen, to command it, under the orders of General Zajączek.

Before crossing the Rhine, General Zajączek harangued the soldiers and had them swear an oath in the Polish manner; that is to say, by swearing on the Bible, and in raising a finger in the air; however, this ceremony caused the general to see that his troops still thought themselves prisoners of war. The Polish officers, themselves, stated that their soldiers could not believe themselves free of captivity, except that they were armed. Otherwise, under various reports, it was very inconvenient that the Legion cross Germany without being armed, that the soldiers would be insulted everywhere they passed, and that they would not have any means of making themselves respected.

General Zajączek thought himself obliged, as commander, to support the demand of his soldiers, and, on his side, Maréchal Kellermann, seeing the danger of arming a mass of prisoners of war, was not inspired with confidence. Finally, the old maréchal decided to give them the poor German weapons they found in Mainz.

The requests of the officers and the troop did not limit themselves to asking for arms. They were heard crying in the ranks, in Polish, that a musket without cartridges was worth less than a stick and that they should be given cartridges.

The troops all lacked shoes. Shoes were placed in a heap at the Mainz fortress the moment when the Legion was crossing the Rhine. General Zajączek held as many of these shoes as he could himself to assure that soldier took more than one pair, and, as the men still wore Prussian uniforms, to assure that no disorders erupted, the maréchal got rid of some of them before they were provided of the papers they needed to be paid while on the road.

It was immediately after the departure that I arrived at my post in Mainz. I was alone and without a horse. I learned that the troops had only left two days earlier and that General Zajączek, according to the orders he received, quickly departed to move to Poland. I was informed that he had given command of the entire corps to Chef de bataillon Vanrosen, while he awaited my arrival. This commander was the only one with the corps. No one other than Vanrosen was named during this time, because the composition of the troops, as well as the orders given on its organization, stated that the superior officers should be both Polish and French.

To that point, it had not been difficult to find soldiers from among the Prussian prisoners of war, since these prisoners, in entering into the Legion, were assured of returning to Germany. Also, there was not a single soldier who did not pretend to be born Polish. Unfortunately, a large number of these were born in Germany, and had later learned to speak Polish while they were in garrison in Posen, and it was easy to fool the recruiters, who questioned them to determine their birthplace.

The difficult work consisted of organizing the Legion as it passed through Germany. It was, furthermore, necessary for purposes of discipline, to find officers and non-commissioned officers, and recruiting while on the march, to bring the Legion to full strength, as well as replacing deserters out of necessity because of all the German soldiers, who had pretended to be Polish, had formed the plan to march with the corps only until they were near their homelands. It was certain that this group would be mistreated by the inhabitants of Germany, who would reproach them for taking up arms against their sovereign, and who would be called upon to desert.

I met, in Mainz, and before I arrived there, many soldiers with whom I was acquainted, had a bad opinion of the events which a similar organization might have dealt with. It was not possible, according to these soldiers, that the French would provide a useful portion from these troops. They said, "The thing that one had to fear most were revolts. These soldiers are capable of killing their officers and returning to Germany in groups to organize a new enemy army behind the Emperor's army." Despite these opinions, I resigned myself to anything that might come my way. I had a strong opinion of my zeal, my courage, and my capacity.

In Mainz, I met a Lithuanian officer, named Hoppen. He was an ardent Polish patriot, who spoke French and several other languages, having served for a long time in the Army of Italy, and who was covered with wounds. General Zajączek had employed him in recruiting at Mainz. He brought me up to date on the Legion, from all my questions. I perceived that he enjoyed the confidence of the Polish officers and the same with the troops. My candor and my loyalty were pleasing to him. He showed himself attached to me. I took him in my carriage and joined the tail of the column. I had taken care to keep my uniform, of the 100th Line Regiment, to inspire more confidence in the soldiers.

From then on, Hoppen became my interpreter and I believe, still, that as a result of him, I obtained the success I did as I crossed Germany. (I only speak of recruitment and organization.) Also, I recommended him, most particularly, to Prince Poniatowski in 1808, before leaving the Polish Army.

The first days of the march were not remarkable, except for a revolt of the Legion in Hanau. It was occasioned by the magistrate and the commander of arms of this city, who wanted to lodge the soldiers of the Legion in churches, as if they were still prisoners of war. This event occurred on 3 or 4 December 1806. I had not yet joined the corps and Chef de Bataillon Vanrosen, who had replaced me, had not dared to resist the commander of arms.

I gave orders, on several days to the effect that I advanced towards the head of the Legion, to choose a certain number of sergeants and corporals from among the men in each company. I gave a letter to some Polish officers, orders which would allow them to acquire vehicles or horses to arrive in the principal cities of the countries where the Legion would pass, in order to gather, in the colleges, all the young people of their nation who wished to follow us. I needed to make these young men officers and to let the troops see that we were truly taking them to Poland. I had rapidly read the history of the kingdom. I did not cease to tell the Polish officers, who knew French, of the conquests and acts of arms that still honored their nation.

Though these provisional officers had never served, it was not appropriate to consider them as useless. To the contrary, they were more ardent and active than those that had seen service, especially when they acted to arrange accommodations for their troops, to distribute food to them, and to assemble them. I did not require another service [of them] while on the road. The officers designated their sergeants, corporals, and the portions of the troops began to take the form of companies. We began to establish the nominal units on which one established the muster, instead of counting the soldiers as one counts sheep, as one had been forced to do before.

In regard to discipline, I constantly recommended gentleness and clemency, because it appeared to me that a certain liberty encouraged the troops and discouraged desertion. The soldiers had left one army in which discipline was rigorously enforced and, if I had not given them a softer régime, they would not have believed in the grand promises I made them as I formed them. I informed each battalion, through my interpreters, and in my presence, that the Legion was the freest corps that had ever existed; that there were only two faults, which I could not pardon. One was to do wrong to one's comrades and the other was to mistreat the inhabitants of the cities and villages.

However, what I had foreseen occurred: Desertion appeared among the troops born in Germany and soon it became a problem, because the number of German soldiers was considerable. I did not see any way of preventing it, because the German soldiers were desperate to return to their families instead of remaining in the Legion, and all my thoughts turned towards the means I could employ to replace the losses I suffered through desertion.

I surrounded myself with Polish officers who had my confidence and I described to them that the founding of the Legion was in our hands. It was not difficult for me to convince them that, if they did not employ extraordinary means, the officers would remain at the head of a handful of unhappy Polish soldiers. I consulted them on the resolution I wished to take with regards to recruiting, which was my authority, in the columns of prisoners of war that we encountered, or which passed close by us. They applauded this measure and promised to support me. Then, I explained to them that it was necessary to only admit prisoners of war into the Legion that were of Polish birth. This extreme order could compromise me, because it was necessary to raise the Legion above the determined strength and, if after the recruitment, it was subject to some act of insubordination, at the instigation of Prussian governmental agents, who could easily gain entry into the ranks, I could be blamed and possibly punished. However, I knew the superior officers and staff officers, who were attached to Gen. Henry Wolodkowicz, who had not been able to form a second Legion, and I showed that I had more troops than that numerous group.

The commanders of the prisoner columns refused to allow me to recruit without an express order, alleging, as a motive for their refusal, that they were responsible for the men who had been confided in their care and that I did not have the right to give them an order. I had founded my rights on the obligation where, as commander of the Legion, I was obliged to maintain it at full strength. However, as it was impossible for them to refuse me, I gave them, or made them give me, receipts indicating the men I had seized.

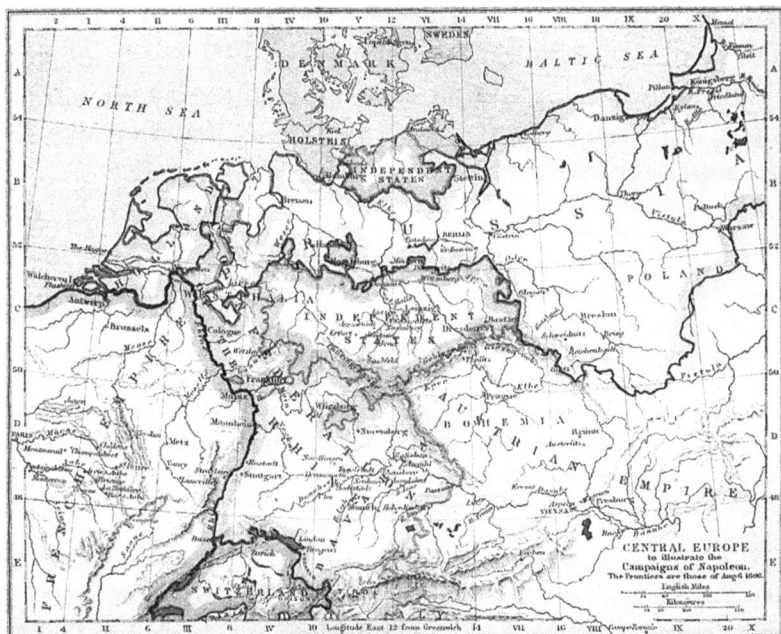

Central Europe 1806

Before arriving in Leipzig, a large number of soldiers had bare feet. There was continual rain. The troops were more and more miserable at their lodgings, because they continually grew in number. The soldiers were demoralized, and I could only get them to march by promising them that, in the first city where there were magazines, they would be provided with everything necessary. It was in these circumstances that Hoppen and the other Polish officers rendered me the greatest services.

The arrival of the Legion at Leipzig was remarkable. The companies were lodged in the surrounding villages. Général de brigade René[4], who was the governor of this region, informed me that a general sent by the Emperor would arrive shortly to examine the Legion. I went to find him. It was General Lecamus.[5]

At this time (I believe it was 20 December) I had so augmented the Legion, despite desertion, that it had risen

[4] Jean-Gaspard-Pascal R, général de brigade 14 December 1801.
[5] Jean Baron Le Camus de Moulignon, général de brigade 1 March 1806.

to more than 10,000 men, according to the states that were sent to me daily by the commanders of the companies, and without including the soldiers who had remained behind for a long time. Until then, I could not have known its strength, because, on the one hand, the recruitment and the desertion had been continual, and on the other, the various columns had constantly occupied from 25 to 30 leagues of the road.

It was through General Lecamus that I was informed of the disturbances to which the Legion had been given. He was surprised to learn, from me, that its force had only risen to 10,000 men, as he had anticipated 30,000, at least. The commanders of arms and the governors had given him reports of removals of prisoners I had made. The Grande Armée had not left troops in either Prussia nor Saxony, so it was feared that one might see the return of part of the Prussian Army into the states of this sovereign under the name of the Legion. The Emperor had not wanted to give faith to all the extravagances that fear had suggested to the commanders of arms. Napoleon had wanted to determine the truth before this suspected army arrived in the Prussian states, and, to this end, he had dispatched a general from the Imperial headquarters. This general had traveled the post route, without stopping, since he left Warsaw until he had met the Legion.

The Legion remained in Leipzig five or six days, while General Lecamus held a review of the various portions of troops that represented the companies, and, in effect, that were brought into the city. This period allowed me the time to begin organizing them into companies and to establish the nominal controls. I prepared a large table that presented, numerically, by company and battalion, the organization that I could outline. General Lecamus provided shoes to all the soldiers that lacked them. He provided me with 5,000 or 6,000 sentry's greatcoats, which, with the 1,200 that General Zajączek had obtained in Mainz, completed the number that was necessary for the first three battalions. I could not obtain any for the fourth battalion, nor for the large detachment of recruits that were coming.

There was, at that moment, in Leipzig, an immense quantity of cloth and wool of all qualities and all colors. I was told that the wool belonged to an English merchant. It had been put under the control of Count Wilmanzy. Working with him was a certain André de la Lozère, who filled the functions of a commissary of war, and whom I had a dispute relative to the material I had received, and another man, named Courpon, who was the guard of this rich magazine. It was this magazine from which cloth was drawn to clothe all the officers of the army, at a rate of six aunes per individual. However, the officers of the Légion du Nord had not participated in this distribution. André and Courpon were prosecuted in 1810 for waste, of which they were accused by the inhabitants of the country, and, to the great surprise of the Minister of War Administration, where I was at that time, they were acquitted.

General Lecamus gave me an order from the Vice Constable to take the Legion to Berlin. I had the 1st Battalion leave on Christmas day. The others left on successive days, one per day.

The day the 4th Battalion left, I left Leipzig, but I took the post, as much to watch the march as to give the necessary orders to all the battalions. It was necessary, therefore, that I join the head of the column.

Arriving in Wittenberg, a happy chance occurred, where one refused to give me post horses without authorization of the commander of the fortress. I went to find him (it was Adjutant Commandant Nivet). He gave me an order from the Prince Vice Constable, that after this order, I was to lead the Legion to Magdeburg. Without this incident, I would have continued my march on Berlin.

I arrived in Magdeburg during the night of 31 December/1 January, and I had the gates opened only with difficulty. The first battalion arrived in that city during the night of 31 December. General Elbé, who governed this region, had given it orders to continue its march on Berlin and the battalion was lodged in the villages a few leagues from the city.

At daybreak, I went to visit General Elbé.[6] I showed him my orders that said the Legion was to remain in Magdeburg. His angry gestures and the tapping of his foot astonished me. He called Chef de bataillon Forgues, commander of the fortress, who was in a neighboring chamber, and he ordered him to lead me to prison. It was difficult to find an officer more determined than I was, then, to brave all the most unfortunate events. For a long time I had sacrificed my life, without meeting the enemy, because the type of troops that I commanded required this.

The efforts that I made to remain silent choked me. I said to him, in a firm voice, "I see that it is serious since you threaten me. You take me, perhaps, as a bad, reformed officer, without character, without honor, without means. But you fool yourself and you will soon see the proof. Before obeying your order, I would have to put command of the Legion in the hands of a superior officer. I warn you, that there is only one with me, and that he cannot relieve himself of this burden. I have no need to be led to prison. If my two epaulettes are sufficient guarantee for you, I can, at least, boldly declare to you that I have as much honor as you. Be careful, if you doubt it. I ask only that my troops not be informed of my imprisonment. You will quickly be better informed on my account. My conduct merits praise, and not punishment. Perhaps the justice that you will soon be obliged to render me will be too late. You cannot reproach me for taking an inconvenient tone, since you have forgotten the respect you owe my rank."

[6]Jean-Baptiste Elbé (1758-1812), the future hero of the Berezina, was governor of Magdeburg at this time.

The general then ordered me to house arrest in my lodgings and I left without saying a word. I only saluted. However, this was beyond what a soldier should show who finds himself exempt from reproaches.

The general had wished that I arrived in Magdeburg with my first battalion, and he supposed that it was from pleasure or from negligence that I had entered the city during the night. He forgot, in this circumstance, the first and most important of all the principals of good discipline, that of questioning an officer that one thinks culpable before condemning him. What is much more surprising is that General Elbé enjoyed the most brilliant reputation among the French troops and the same in the armies of Europe. He was, perhaps, the first general of artillery on the continent, and, if the Emperor had placed him as a governor in the rear of the army, in 1806, it was to punish him for being with the Army of the Rhine and a friend and confidante of Pichegru and Moreau.[7]

---

[7]Editor: Moreau, Jean Victor Marie (1763-1813): Moreau was one of the major generals of the Revolution. He commanded many armies and, in 1800, in a parallel campaign to Napoleon's Marengo campaign, he won the major battle of Hohenlinden. His reputation as a soldier rivaled that of Napoleon.

Shortly before Hohenlinden, Moreau had married Mlle. Hullot, a creole of Josephine's circle and a very ambitious woman. She gained complete ascendancy over him. She collected around her those who were discontented with Napoleon's rise to the Imperial dignity. Napoleon was well aware of the "club Moreau" and very annoyed by it. However, Napoleon knew that Moreau might be willing to become a military dictator, but also knew that he would not support a restoration of Louis XVIII. Irrespective, Moreau represented a threat to Napoleon, so he was arrested, tried, and his sentence of imprisonment was commuted by Napoleon, as a sign of leniency, to banishment. Moreau moved to Morrisville, New Jersey, and lived there until 1812, when he learned of the disasters met by the Grande Armée in Russia. No doubt at the instigation of his wife, he opened up negotiations with Bernadotte, who introduced him to Czar Alexander. In the hope of re-establishing a popular government in France, Moreau provided advice to the Allies on how to conduct the war. Moreau resumed his uniform and joined the Allied armies in the field against France, accompanying them in the fall of 1813. While on the fields by Lahn, during the battle of Dresden on 27 August 1813, he was struck by a cannonball and mortally wounded. He died during the night of 1-2 September 1813. Popular rumor had it that Napoleon, himself, had laid the cannon that killed Moreau, but there is no evidence of this. Moreau is buried in St. Petersburg.

Pichegru, Jean Charles (1761-1804): Pichegru was a professional soldier

These last considerations determined me, in Magdeburg, to contain as much as it was possible for me to do so. If I had a parallel altercation with a mediocre general of little esteem, I would have taken a different tone.

Arriving at my lodging, I wrote General Elbé a letter explaining to him why I had arrived in Magdeburg after my first battalion. The expressions that served me, in speaking to him of my men, had caused him to fear my men. I informed him that I would leave the army and, if he was shocked to see a secondary governor in the rear of the Army of the Rhine, he was no less shocked to find me at the head of a semi-savage corps. He subsequently had me released from my arrest. I went to pay him a visit, where he welcomed me and invited me to dinner.

I knew that the general had learned, before my arrival at Magdeburg, that I had recruited prisoners of war who had served as part of the [pre-war Prussian] garrison of this city, and that he feared that the soldiers of the Legion, encouraged by the inhabitants, would make themselves

---

before the Revolution. He was initially a Jacobin and this opened up a very rapid series of promotions and command. His greatest victory was the conquest of Holland in 1795. He commanded several of the major field armies and there is considerable evidence that, in late 1795, he was conspiring with the Royalists to overthrow the Republican government, but it is unclear if he was looking to re-establish the Bourbons, or possibly establish himself as ruler of France.

Pichegru left the army and was elected to the Council of Five Hundred on 12 April 1797. Pichegru rose to the post of President of the Council. He was a Royalist leader and planned a coup d'état, but on 18 Fructidor, he was proscribed and arrested on 4 September 1797 and deported to Cayenne, Guiana. Pichegru escaped in June 1798 and reached Surinam, taking refuge in London. He served on Korsakov's staff in the 1799 campaign in Holland, fighting against the Republican French armies. Pichegru took part in the Cadoudal-Pichegru Conspiracy and secretly went to Paris. Pichegru was taken by the police on 28 February 1804 after being revealed by one of his former officers, Le Blanc, with whom he had sought refuge. He was imprisoned in the Temple and found strangled to death on 4 April 1804.

Napoleon was very unsteady on his throne and any potential rival was dealt with harshly. His three biggest rivals were Pichegru, Moreau, and Bernadotte. There was, within his concerns over his position, a considerable amount of guilt by association, so if Elbé had served with the first two of his great rivals, he was very suspect until he proved himself differently.

masters of the city. The French only had two provisional regiments of conscripts, like the Legion, occupying the city and they were only in transit, who remained in the city during the entire period the Legion was there. It is certain that these two regiments were far from being able to defend against the Legion, and, in case the Legion should revolt, it would have been prudent to send the two regiments of conscripts into the citadel to cannonade the city. The Emperor almost always neglected the rear of his army. Magdeburg, Spandau, and Wittenberg were, however, fortresses in which he left garrisons.

General Elbé recalled the first battalion to the city and the three other battalions arrived successively.

The Legion remained in Magdeburg for ten days, without my first knowing the reason for this unusual sojourn. But, after several days I knew that I was to give General Clarke, Governor General of Prussia, the time to arrange, in Berlin, the measures without which it would have been totally impossible to send the Legion into the field.

However, I mistrusted events that could follow the quarrel that I had with General Elbé, because this general had made no report of it and the Emperor could still be informed of it by his secret police. I addressed, to the Prince Vice Constable, a report, in which I informed him of what had happened, and I sent it as quickly as I could, so that it would arrive before any report from General Elbé. I knew well that the assistant governor was under the orders of General Clarke and that it was, without a doubt, to him that he would send his report. However, as my conduct merited praise and not reproaches, I wanted the Major General of the Grande Armée know of it.

I had lost around 900 men through desertion between Leipzig and Magdeburg, but a detachment of 700+ Prussian prisoners of war arrived in this last city, sent from Mainz, by General Kellermann, and led by Chef de Bataillon Roumette. I received, with this detachment, a number of officers, of whom I had great need.

I had known Chef de bataillon Roumette as the colonel of the 10ᵗʰ Line Infantry Demi-brigade, during the blockade of Mainz. It appeared to me that he had conducted this detachment well and I gave him command of the 3ʳᵈ Battalion, without any authorization beyond the order he brought with him.

Governor General Clarke gave me orders to go to Spandau as the Legion left Magdeburg, battalion by battalion, around 10 January 1807.

Chef de bataillon Roumette had brought, with his detachment, a Polish captain named Grabinski, against whom he presented to me a complaint supported by proof. This captain had continually agitated the detachment to rise up against Commandant Roumette and the French officers had been obliged to forcibly hold him away from the troops during the march, with military threats. Grabinski was also accused, by all the officers, of having attempted to provoke the desertion of 500 or 600 men, while crossing Hesse, and where the detachment had to act against a popular uprising.

As Captain Grabinski was Polish and very familiar with the soldiers, I judged that I should not reprimand him, and still less, have him arrested. I had him come to me and I spoke to him in pleasant terms, making every effort to gain his confidence. I spoke to him of the Polish officers with whom I was happy and the promises he made me left me nothing to desire. He concerned himself, at the moment, to arrive at Spandau in good order, and nothing besides.

The day that the 2ⁿᵈ Battalion was to depart, I gave orders to Lieutenant Guillemin de Vaivre, a returned émigré, with whom I had observed had a good military spirit, a great deal of character, and ability. The captains appeared too mediocre to me to command battalions as numerous as in strong regiments and whose composition required a lot of skill on the part of those that commanded them.

At the time the 2nd Battalion assembled to depart, Lieutenant de Vaivre, who commanded it, came to tell me that Captain Grabinski appeared on a horse belonging to M. Grosourdy de Saint Pierre, Knight of Malta, who had been sent to the Legion as a captain, and that he rode in front of the battalion, speaking to the soldiers in Polish. Vaivre said that a servant of his had told him that Grabinski had urged the soldiers to rise up against him, and that he exhorted them to only recognize him, Grabinski, as their commander. De Vaivre had lunch and asked me for permission to blow out Grabinski' s brains.

I spoke confidentially with de Vaivre and authorized him to shoot Grabinski, if he presented himself, during the march, before the battalion and attempted to incite them to revolt. I then moved to the fortress, and I hid with Grabinski, feigning ignorance of what I had been told. I expressed to him that M. de Saint-Pierre had charged him with his horses and his caisson, and similar military equipment necessarily that kept him occupied, and that it was, for this reason, that I had not given him the command of a battalion. I made him understand that, by traveling alone, he would be able to gain the confidence of M. of Saint-Pierre, who, at that moment, was already at Stettin, where he had been sent, and I gave him an order for it, with three or four soldiers he asked of me. I wanted to gain time before we reached Spandau, where I intended to secretly arrest Grabinski, through the intervention of the Governor Général.

During the march from Magdeburg to Spandau, desertion was considerable, because they had been given shoes and great coats at Leipzig. The soldiers had thought, as a result of the promises I had made them, that they would be issued uniforms at Magdeburg. They were very discouraged when they saw superb uniforms given to the Grand Duchy of Berg Regiment in Magdeburg, [but that they received nothing.] They knew what was in the magazines of this city and their hopes were disappearing.

Since the month of March of the same year, Prince Murat had been named to rule the Grand Duchy of Berg. It had no army; however, it was to raise one. The expense of this type is enormous, and the prince spent, prodigiously, part of the revenues of the country, which his brother-in-law had given him. He was with the Grande Armée as the commander of the cavalry. The charges his cavalry made against the Prussian cavalry had been brilliantly successful and his brother-in-law, the Emperor, had given him permission to raise a regiment of infantry at Magdeburg, in order to spare the blood of his subjects. This grand duke had also obtained permission to add French commanders and soldiers to the veteran officers that he had found in his states. He sent a cadre to Magdeburg to recruit there and to fill out the regiment. Geiter, a native of the Palatinate, and a veteran non-commissioned officer of a Swiss regiment in French service, and who was major of the French 15th Légère Regiment, in 1806, was chosen, by the Grand Duke, to be colonel of the regiment, if he raised it. He arrived, in effect, to form the regiment in Magdeburg and it was not difficult to dress it there. The Grand Duke knew more about how to lead the cavalry charges that he made against the Russians and Prussians than forming a regiment. He directed to Düsseldorf enough white cloth for the infantry and enough blue cloth for his cavalry than was necessary to uniform 10,000 men. The cloth found in the German magazines, and which could be used by the Berg troops, was taken along with the other cloth. I had proof of the removal of this material into the Duchy of Berg, where I was sent, in 1812, as the administrator of war, in my capacity of under- inspector of reviews, a position that was given to me upon my return to France in

1809.

I now return to the arrival of the Legion in Spandau. General d'Agoult[8] was sent to Spandau by the Governor General, to receive the Legion, and to prepare the distributions which had to be made to it. He informed me that the General Governor had ordered prepared, at Berlin, breeches, vests, and gaiters, where they also assembled underclothes and shirts. Greatcoats were given the soldiers who did not have them. Felt shakos were found in Prussian magazines, to which were fitted visors that were cut from cases of the shovels, and which were then distributed to the Legion. They were also distributed their major equipment from stocks of new material, replacing the bad equipment that had been given to them in Mainz; finally, new Prussian muskets were issued to all the non-commissioned officers and soldiers. In addition, old swords, which had been carried by the troops of Frederick the Great, were issued to the non-commissioned officers, to the grenadiers, and to the voltigeurs, because there were no sabers to issue them.

General Count d'Agoult was a very handsome man, more than 50 years old, a returned émigré, who had, before his immigration, been a major general in the military household of Louis XVI. He addressed to me, in Berlin, before coming to Spandau, a long instruction, written in his own hand, and of which I understood nothing (I have never seen such gibberish; however, the Count d'Agoult spoke clearly). When he arrived at Spandau, he asked of me if I had executed his orders, and, as his instructions had caused me to doubt he was still in his right mind, I responded that the Revolution had overthrown the ideas in vogue with the soldiers of Louis XVI's army, I did not understand the whole of his instructions, which would surely have been perfectly understood in the Royal Household, and in the Piémont Regiment, in which I had

---

[8] Pierre-Nicolas d'Agoult, général de brigade on 22 December 1800.

served, but that they were unexecutable for the new French infantry, and above all, for a corps of half barbarians that was not yet organized. The general responded to me that he had seen great truth in what I had said that the army had totally abandoned the old methods and that he should learn the new methods.

I should remark, in this situation, that the activity of Governor General Clarke was almost unbelievable. If I had not, myself, made the distributions with General d'Agoult, I would not believe it possible to prepare such a great quantity of equipment and to bring them to Spandau in the space of five days. The Governor had organized the equipment so well that, right after a battalion was uniformed, equipped, and armed, the Governor came to Spandau to review it and to immediately send it to Stettin.

In Spandau, there occurred an incident that displeased me and the Polish officers. In this city and region, the Prince von Isenburg, whose principality is near Frankfurt-am-Main, was recruiting a regiment. General d'Agoult had placed, in his headquarters, an adventurer, supposedly a Polish captain, by the name of Baron Riéner, and who served d'Agoult as an interpreter. (This rank is commonly taken by intriguers, who sought their fortunes in various armies and spoke many languages.) This supposed officer, under the pretext that the governor had charged him with interrogating soldiers to see if they actually were Polish, drew from the ranks of the Legion a number of soldiers which he delivered to the recruiters of the Prince von Isenburg, who took them under escort by a detachment which had loaded weapons in front of them, and, if I had not stopped this forced recruitment, this Baron Riéner would have stolen a quarter of the Legion. I complained vigorously to General d'Agoult, who took my remonstrance badly, which gave me to believe that I should move away from there. I permitted myself to say that I formally opposed the removal of a single man without a written order of the Governor, and that by admitting that this order was given, I permitted myself to discuss it with him. That said, the general took an authoritarian tone with me, saying: "Your first duty is to obey. Are you the inspector general? Are you aware that I can have you arrested?" "General," I responded, "let's forget the arrest, as it is too dangerous. It would be impossible if I was in the mood. You know, as well as I that one does not arrest a commander who is at the head of 9,000 soldiers, who does not recognize you as his superior, and, above all, who possesses their confidence. Neither you nor the Governor can prove to me that I am in the service of the Prince von Isenburg and that it is for this little sovereign that I have recruited the Legion. I am more interested than you that there are only Polish men in my command, and I must report to my Polish officers that you have certainly been misled by this adventurer."

The seizure of men had already occurred in the first battalion, and I did not want them to touch the three others. My men were armed. It occupied a fortress that it knew well, and an insurrection was still to be feared as the inhabitants agitated my soldiers, telling them that they had done wrong to arm themselves against their legitimate sovereign. Constantly, instead of directing my complaints to the Governor General, I took part in threatening the supposed Baron de Riéner by the Polish officers that had my confidence and this measure produced the effect that I hoped, because he found no German soldiers in the 2nd, 3rd, or 4th Battalions. I was, myself, more interested that he found none. I knew that the Legion had to cross the Oder and that it would find itself constantly in the middle of von Schill's partisans, dressed as locals, spread through all the cities and villages where my troops would pass, and that they would not fail to torment my soldiers and call on them to desert.

The Governor General was surprised at the height of the soldiers in the Legion. In one company of grenadiers, there were 37 men who were at least six feet tall or more. These large men had come from the King of Prussia's Guard.[9] The fusiliers were at least five feet three, or four, inches tall.[10]

Before leaving Spandau, I recounted to General d'Agoult the bad conduct of Captain Grabinski, of whom I have already spoken. I told him the reasons why it was dangerous to arrest him, and the Governor General had him imprisoned the moment the Legion started to march out.

It was at Spandau that I was given a number of officers from the 2nd Legion, which was to be formed at Nurenburg, along with 43 soldiers, dressed as beggars and who brought with them about 80 women. The officers were all very young, and by their situation, looked more like a crowd of scoundrels. I listened to them quietly and, as they had no doubt of admission of their troops, each of them designated to me the soldiers that should be incorporated into the company, or where they should be placed. Such was the cadre of the 2nd Legion of the North, which General Henry Wolodkowicz had to recruit and organize. I quickly saw that the 43 soldiers were Germans and that they were worth less than those that I had refused during my march. I told the officers that I would receive them with pleasure if they wished to take up the functions that I gave them in my companies, but that I did not want their harem; that the recruiters of the Prince of Isenburg were, perhaps, less difficult than I, and that they could speak to them. The officers did not wish to leave their

---

[9] Editor: The Prussian Giant Grenadiers had been disbanded early in the reign of Frederick the Great. Though it is not impossible that taller men were still selectively placed in the Prussian Guard, it is very possible that this was quite possibly apocryphal.

[10] Editor: Measurement systems are a confusing issue at this time, which is why the metric system became so popular. If these are French "feet" from before the institution of the metric system, 6 "pieds" is approximately 6 feet 6 inches tall (about 2 meters). The height of the fusiliers would be 5 feet, 8-9 inches tall (1.75 – 1.77 m).

troops, and at the end of this period, I will no longer speak of the 2nd Legion.

The Legion arrived in Stettin [poorly] dressed, as I have said earlier, but it was necessary that I recall, here, our situation, in order to let one see that the French soldiers would not resist going into bivouac during such rigorous winter, if they were so lightly clothed.

Their great coats were of such poor material, crossed and loose, that one could almost see through them, and were brown in color.

Their rounded waistcoats were made of a crossed and thin material that would not have lasted a year in the Prussian Army.

Their breeches were of the same material as the waistcoat but were cut straight and were unlined.

The gaiters were made with the same cloth as tents[11], but strong, with buttons of black leather.

Their shoes were drawn from Prussian magazines. As a result, they were of bad quality.

Their shirts were also made with the same material as the tents, and, consequently, of the same quality as the gaiters.

Their stockings were of blue wool but could only be worn by the soldiers for a few days before they tore.

The neck stocks were of black leather and were also drawn from the magazines of the Prussian Army. No one had dared to give them to the French soldiers.

Their shakos were of bad felt, which soon turned white, and to which were attached visors that had been cut from the cases of shovels.

The movement of the Legion to Stettin occurred in the last days of January 1807. General Duphot[12], who I

[11] Editor: It is surprising that the author is so clear that this is the same material used to make tents. Normally, the gaiters are described as having been made of canvas.

[12] Duphot is an anagram for Puthod, whose name the author disguised, as he does with the names of many officers further on. Jacques-Joseph Marie baron, later Count Puthod (1769-1837), joined the army in 1785, became a captain in the Colonel Infantry Regiment and distinguished himself in 1792 during the siege of Lille. He

found in Stettin, told me that the Legion was under his orders and that it formed the infantry of his brigade. I went to him to study and to discover, by conversation, his military habits. I found him dressed in a hussar's dolman that was trimmed with a beautiful fur and embroidered on all seams and edges. I must have a bad opinion of an infantry officer dressed in this manner. His gestures, his manners, his light tone, and his words struck me as being in perfect harmony with his costume.

[Puthod] had with him young officers, who made, under his supervision, the silver braid he wanted to buy for the rank insignia for the non-commissioned officers, as well as to embroider their shakos and to trim the drum major's uniform. He added to this expense that necessary to provide the cords for the shakos and to purchase 36 pennants of colored cloth to be used by the 36 companies of the Legion.

I was shown the state and I was asked my opinion of the materials he was going to purchase. I responded they were the most beautiful necessary and I asked the price he was paying for them. He responded, "Ah! Ah! I know well that you have 50,000 francs in your chest, and it is there to be spent for the service of the Emperor!" I responded, "The 50,000 francs were given to me to purchase linens and stockings for the first battalion; General Zajączek had already taken too much from this first allocation, for objects that were unknown to him, and the remainder will only leave enough in the chest for a good ensign!" In short, a long discussion began between the general and me. This discussion completely confirmed, the opinion that I had formed of him. I saw, clearly, that he was one of those generals who had never commanded troops and who owed their advancement to the storms of the Revolution. I recalled this particularity because, later on, General

served under Moreau as a général de brigade, having been promoted to that rank on 19 October 1790, and finally became a général de division in 1808. Editor: Henceforth, [Puthod] has been inserted into the text to clarify this.

[Puthod] presumed to replace General Zajączek in command of the Legion without having received such orders. I found myself with him, one day, when he described his past escapades. It didn't escape me that he took pleasure to say that he didn't have any morals, and, from there, I presumed that he would show little or no bravery before the enemy, because bravery is a virtue, and it was very extraordinary to find

Legion Nord - Grenadier, Fusilier and Voltiguer *by JOB*

it in one who says that all he has is vices. One will see, by what I will relate, that this conjecture was only too well founded.

On 5 February, at 11:00 p.m., General [Puthod] sent me an order to distribute, four days of bread and meat, and cartridges during the night so that they would be ready to leave the next morning at 7:00 a.m. They were to march out with the Baden cavalry and artillery attached to his brigade. The Baden artillery consisted of five 4pdr guns and a 6-inch howitzer. I was advised that it was attached to my Legion, so I responded that I would provide to the men and horses the rations that was their due.

I went to General [Puthod] and spoke to him in these terms: "Although we are not in front of the enemy, I don't doubt that you had good reasons not to have given me the order to distribute rations before last night, but there are a thousand precautions to take to give to the Legion an order like that which I received from you. I advise you, finally, that you don't blame me for the mess that its literal execution could entail."

What I anticipated occurred. At daybreak, one saw nothing but soldiers from the Legion in the street who were intoxicated, sleeping on the ground or staggering about, throwing their cartridge boxes, breaking their muskets on the pavement, swearing or giving other proofs of the most complete mutiny. One saw the same tumult on the esplanade where the battalions were to assemble. I then engaged all the officers to walk confidently among the soldiers on the esplanade and in the streets, maintaining a composure equally in good humor as well as serious, and to keep themselves from being surprised; to joke with them that they should to their duty, and to make all the efforts to inspire their confidence by removing from them any fear of punishment.

I went to General [Puthod] to inform him of what had occurred. I found him, certainly, calmer than he had been the night before during our interview, and, on this occasion, he did not discredit my observations. He told me that it was General Dermans[13], commanding a division consisting of the Baden troops and the Legion, who had given the order, and to whom General [Puthod] was to conform. I saw later that General [Ménard] had much more judgment and finesse than General [Puthod].

---

[13]No French general by this name existed. We have to assume that this general, whose name has been changed, is a foreign officer. Editor: An examination of orders of battle for 1807 show the following officers commanding formations with Puthod: Dufour, Ménard, Crown Prince of Baden, and Ménardn. It is probably Ménard, who commanded the Baden troops, that is being spoke about here. Sauzey, *Les Allemands sous les aigles française*, Vol II, *Le Contingent Badois*, pg 11-13 indicates he commanded until sometime after 31 March, when the Crown Prince of Baden arrived to assume command. Ménard's name is specifically mentioned later, so it was not him. This leaves only Ménard, whose name will be substituted, henceforth, in the text.

After my invitation, General [Puthod] came out to see what had happened. He went to the esplanade and then returned to his lodging.

What could one do against the Legion and the convalescent depot, which formed the garrison of Stettin? General Thouvenot, the governor, was anxious. A considerable number of soldiers from the Legion, in throwing down their muskets, breaking them on the pavement, declared that they would not serve against the King of Prussia. These demonstrations resulted from the instigations of the troops of [von] Schill and the inhabitants of Stettin. This city had been occupied, since 29 October 1806, by a brigade of hussars commanded by General Lasalle[14], as a result of a shameful incident for the Prussian Army that contrasted markedly with Colburg, Danzig (Gdansk), and Graudentz, which did not follow its example. At the time of which I speak, the beginning of February, the partisan officer, von Schill, had already taken, in a village near Stettin, General Victor[15], who had been lured to that village by a lady of Stettin for a rendezvous. This was not the General Victor that the Prussians had wanted to capture. His reputation was not sufficiently brilliant for him to merit such an honor, but he became commander of the siege of Danzig, and he carried day and night, on his person, instructions that the Prussian generals sought to know.

I now return to the Legion. It is beyond doubt that the partisan, von Schill, was, himself, in Stettin, at the moment when the Legion mutinied, and that this daring commander had questioned my soldiers for five days, and had sought, through this mutiny, to turn their arms against

---

[14]Antoine-Charles-Louis Collinet de Lasalle (1775-1809), the celebrated cavalry general, was killed at Wagram.

[15]General Victor, future Duc de Bellune, was, in effect, captured near Stettin, with his carriage, his aide-de-camp, and his servant, by 23 Prussian partisan light cavalry, as he moved to rejoin the French Army besieging Danzig. The French accounts are very temperate in their relations of this incident, while German accounts, by contrast, make great noise about it. Victor was quickly exchanged, by order of Napoleon.

the French depots which formed the Stettin garrison.

Towards 10:00 a.m., I thought I saw that the mutineers were less inflamed. I charged the Polish officers to move through the streets and the esplanade, to say to the soldiers that French troops were going to enter the city and put it in order.

This measure produced some effect. It was possible to begin assembling them on the esplanade. The soldiers gathered up their muskets and cartridge boxes that they had thrown away. Generals [Ménardn] and [Puthod] finally dared to appear on the esplanade. I ordered the muster of the 1st Battalion, alone, and then ordered out those that gathered there. The same operation was then done for the remaining three battalions. The soldiers who, after the departure of the four battalions, were still found in the streets, became afraid (and they had much to fear), and marched out alone, reaching the bridge on the Oder. The order had been given for the Legion to be lodged at Dahm, two leagues from Stettin, on the far side of the river.

When I saw that there were no more, in city, than a few stragglers, I set out for Dahm. The stragglers, in great numbers, marched independently. There were many who marched on the right and left of the road, who appeared be marching on Dahm. I was with Lebrun, quartermaster-treasurer of the Legion. The scattered men fired their muskets continuously at us, as far as four leagues from Dahm. We heard the whistling of the balls, around us, as if we were in a skirmish fight.

Upon arrival at Dahm, I found great confusion and disorder. There were, nonetheless, a large number of soldiers that the officers had in ranks, but they fired their muskets. I was shot at point-blank range in cheek, so to say, by a soldier that was drunk, or who feigned to be. The troops were amused until nightfall and obscurity obliged everyone to enter the houses.

I have never doubted that it was one of von Schill's soldiers who shot me in the cheek. I quickly threw my horse against him; I grabbed his musket and pulled it from his hands. I jumped down from my horse and I gave the soldier a fatal blow to the head. The multitude around him made no movement. I did not remount my horse because, by remaining on foot, no one could fire on me without also firing on everyone else.

The Battle of Tczew by *Juliusz Kossak*

The following day, I had the Legion march out, with much difficulty, towards Stargard, a city six leagues from Dahm. The muster of that morning indicated a loss of 900 men.

Several days after our departure from Stettin, General [Ménard], filling the functions of a division commander, wished to hold a review of the Legion, so he might examine it. It was placed on the back of a hill, in the snow, and in column by battalion, at full intervals. Some men were entrenched on the hill very near there and firing their muskets at us, or on the generals or commanders, who were all mounted. I judged the direction of the fire by the whistling of the balls. General [Ménard], who was at my side, said, "We are in bad company, major!" I moved my horse against his and I moved alongside him to mask him. I sent Hoppen to the hill to reconnoiter the skirmishers, who were hidden behind a small mound. He soon returned and gave us a satisfactory response. I have the confidence of the troops. I didn't doubt at all, and I always believed that the skirmishers, of which I speak, were von Schill's soldier's, who were wearing greatcoats belonging to the Legion. Hoppen has admitted to me, since, that he had spoken to them, in German, politely, so as to not be shot at. General [Puthod] and his aides-de-camp had retired towards the 4th Battalion, which was further from the skirmishers. A very deep ravine, and which was full of snow, prevented a closer examination of the skirmishers.

The Legion continued its path in a country that appeared to be deserted. We found neither food nor lodging there. It was extremely cold. The ground was covered with much snow and the troops were fed as had been planned.

Soon, one only marched by encampments. I bivouacked my troops by having them enter some buildings to defend themselves, something which is never seen in a parallel case. The truth was, we marched in the middle of von Schill's army, but which was not to be seen, and had there really been a danger, one could have guarded oneself militarily, in the cities or villages, during the night. I could no longer doubt that I was under the command of a bad general, but I was armed with patience. To that point, von Schill's troops, those of the Danzig and Colburg garrisons, which watched the French columns, had not dared to attack the Legion, because it was too numerous and, it is necessary to add, that the Baden troops marched in front of us.

The commanders and the officers of the Polish troops, which we encountered, told us that it was General [Puthod] who had bivouacked the Legion with his own authority; that we were the only troops which marched in this manner, and he was still angry after the insurrection at Stettin. The encampments we made cost us many soldiers. The officers spoke to me, continually, and I decided to submit my observations to General [Puthod] on this subject. But they were not better welcomed than those I had given him in Stettin. I pitied, above all, those who had to march 12 to 15 leagues per day because they did not pass through the cities and the villages that his aide-de-camp, Denarbe, had visited earlier. It was the guides that this miserable aide-de-camp brought, who caused us to make detours.

Towards 15 February, I believe, and the day following my quarrel with the bad general under whose orders I served, we bivouacked in a swamp, at the edge of a forest, without straw, while villages were all around that location. The soldiers had seen them. The generals and the Polish commanders, with many of the officers, had come to see the Legion the day before and had spoken extensively with the soldiers. I remained on foot through most of the night, to distribute the meat which had been obtained from the villages. I was most unfortunate not to have anyone in which I could place my confidence, and I directed the distributions myself.

The following day, from daybreak, I mounted my horse to go, myself, to receive the muster of the companies. I took notes with my pencil, and I found that I was missing more than 1,500 men from the muster. I still hold the situation report that I had the quartermaster draw up as a result of this muster.

That same day, during the march, the general asked me what could be done to prevent this desertion. I did not respond to him initially, out of fear of being carried away. He insisted, and I limited myself to saying, "Do you recall what I told you the other day, if you did not want to arrive before Danzig with only a few officers?"

I have no doubt that the generals and the Polish officers, all of whom were owners of serfs, and obliged to furnish their contingent to the [Polish] national army, which was being raised, desired that my entire Legion desert and join their troops in order to be returned to their lands, as many serfs as the Legion furnished them with soldiers. I had to believe, after this, that my 1,500 deserters were mixed amidst the Polish troops. But the bivouacs were no less the cause of this loss, since, from the day that they ceased, the desertion also stopped.

The Polish officers had proposed to me, several times, that I relieve the Legion of a general, under whose orders it would evidently melt. However, I was gifted with a strong conscience that sustained itself in the most stressful circumstances, and, to the consideration of the one of which I speak, I believed I could not divert the Poles from their project than by frightening them with the continuation of what was meditated: "There is no assassin in the Legion," my Polish officers told me, "and, if we knew one, we would not suffer him here. However, one man can, suddenly, find himself out of a state to continue the campaign, without ceasing to carry himself well." I knew, since, that someone had proposed to cut his Achilles tendon. This general had Captain Molosan[16] and Lieutenant Denarbre as his aides-de-camp. [Salomon] spoke better German than Denarbre, but he was not the one that went through the cities and villages, which surrounded our encampments. I had always thought

---

[16]An anagram of the name Salomon. Editor: Salomon will, henceforth, be inserted in the text.

that [Solomon] was an honest man and I am pleased to render justice to him. The other was of the same mold as his general, and it seemed to me he came from an obscure family, while [Solomon] came from a family that was remarkable by its rank, as it was by the positions it had previously occupied.

The day of 18 February 1807 was a remarkable one for the Legion. I no longer recall if it fought or did not fight against the Prussian troops, with all the non-commissioned officers and soldiers I had left. In truth, they had made me promises, but there were, in the Prussian troops facing us, many soldiers born in Poland who could neither desert nor refuse service, and I feared that a sentiment of patriotism or of compassion would cause my soldiers to make temperament in their favor. I had inspired the confidence of the troops of the Legion, but I feared that I had more of von Schill's soldiers in my companies, and that they had organized, in advance, some new catastrophe.

The Legion arrived in Stargard at 10:00 p.m. I put the Legion in line of battle along the four sides of the fortress, each battalion was with its back to the houses. I sent the four adjutant-majors into a house, where there was a light, and I issued to them my orders relative to the organization of the grand guard that I judged necessary for the night. The artillery was left behind a small river, in front of the entrance to the city, but with a guard of 80 men, in small posts. Five guards of 50 men were placed in front of the city, on the side facing Danzig. These guards were linked together by small posts placed in front of the line that they occupied. General ...... (sic) was lodged in the fortress. General ......[17] (sic) and General Clossmann, of Baden, was lodged on the street to the left which butted up against the fortress and led out of the city on the right

---

[17] Coqueugniot leaves blank the names of Puthod (Duphot) and Ménard (Dermans) here. We will identify them, as much, as possible.

of a small woods. The Baden troops were to arrive during the night, I was told, and several colonels were already in the city.

The magistrates lodged the Baden colonels with me in a house belonging to a Prussian major who was a prisoner of war released on parole, and all our requests to be lodged elsewhere were useless. All the officers of the Legion were lodged in three houses.

General [Ménard] didn't want me to post a special guard at his lodging, according to normal practice, under the pretext that the guard force of the fortress provided him with a sentry and that was all that was necessary. Nonetheless, I obliged him to consent to taking a force of 12 grenadiers, commanded by a sergeant and a corporal into his lodging.

At 11:30 p.m., a heavy burst of musketry was directed at the advanced guards and, principally, towards that on the left. De Pradines, who I had accepted as an officer at Leipzig (he was from the King of Prussia's gendarmes), and who commanded a company of the Legion, was part of the guard at one of the posts, outside the village. He pretended that his post had been forced and he returned to the city with the soldiers who remained with him. The night was dark and only snow was visible. The musketry continually grew in intensity, and I deployed the companies in line before the fortress, as they arrived pell-mell. The officers seconded me. Commandant Roumette pulled together his battalion, after a great deal of activity, and he asked me to which point he was going to fight. I told him that we going to support the posts by detachment; that the main body of the Legion remained in front of the fortress to throw forces at the points where the need arose, and that, if the enemy penetrated into the city, we would be there to receive them. Already, Captains Dubois-Haumont and Vernier had moved, with 300-400 men, to the point where Pradines' guard had been forced. Captain Vernier had already been killed and it was reported that Dubois-Haumont had his thigh broken, which caused me to believe the engagement was serious at that point. I sent the companies everywhere I heard a more active fusillade. Orders were continually being sent, either asking for information on various points or that I be given a report. I waited, in vain, for the arrival of the generals. General [Puthod], lodged in the place of arms, cowered in fear, with his aide-de-camp, and I was to presume that General [Ménard] and General von Claussmann did the same. However, I learned, somewhat later, that the latter had been content to hide themselves in the basement of the château, where General [Ménard] had wanted them to be lodged (this general loved chateaus and bought one in France.

A squadron of mounted troops, wearing white coats, arrived at the fortress during the height of the affair, coming on the side where one of my posts had been forced. I thought that it was part of the Baden cavalry that we awaited, without thinking that it was coming on another side than that where it was expected. I had disturbed the left of my 3rd Battalion to allow these troops to pass. It was so bold as to deploy in line in the middle of the fortress and of the Legion, facing the house of the burgomaster. The horsemen then fired a number of pistol shots. A volley of musketry erupted from the houses all around the place and the Legion made no movement. At the moment when the officers of the Legion went, at my request, to ask the cavalry why it was firing its pistols, it moved to leave the fortress, by the road to the right, on the Danzig side of the city.

The fusillade continued and appeared to close on both flanks, despite the fact that I continued sending reinforcements. De Vaivre, adjutant major, who I had ordered to direct the reinforcing companies and to return to me with a report, arrived at a gallop and told me that he had ridden through a force of Prussian cavalry, which had fired several pistol shots at him and that these troops were not the Baden cavalry I had been expecting. (He didn't misrepresent the pistol shots, because, when it was daylight, he found that his horse was wounded in the mane and that his coat was pierced with a ball.)

Prisoners of war began to arrive, and this was certain proof of the success the detachments had achieved. Belhomme, surgeon aide-major, had commanded a company by himself, because of the lack of officers, on the 17th (he would later give many proofs of his bravery). Dietrich, surgeon sous-aide-major, and a very young man, also commanded a very large company, and comported himself extremely well. I had the prisoners placed in the chambers of the town house. I had no one to treat the wounded, because the surgeons were fighting as commanders of the companies. Surgeon major Dupont was lodged with the general and I thought all the headquarters was in flight, or at least prisoners of war, as I had not seen it and having no news since the arrival of the Legion in the city. I was unconcerned for the artillery, because there had been no action behind the city, where they were located.

At last, the firing having ceased, General [Ménard] appeared, as well as the aides-de-camp of General [Puthod]. These latter had their handkerchiefs around their head and, at first, I thought them wounded. They warned me that the general was going to appear. I didn't answer the question that they directed to me while telling them: "Screw yourselves and tell your general that it is here that he should have been since 11:00 p.m., last night." General [Puthod] informed me that he had seen to the security of the cannon. "I doubt it," I responded to him.

General Ménard, constantly filling the functions of a divisional commander, sent the 1st Battalion of the Legion outside of the city to reconnoiter the vicinity and they did not encounter a single enemy. Soon afterwards, a Baden column arrived, and the divisional commander put himself at their head, to make a sortie would only be ridiculous, after my five grand posts were re-established. The first battalion remained under arms and all the rest of the Legion returned to its lodgings.

My officers were much amused to see the two generals, who had hidden themselves during the affair of the previous night, then put themselves at the head of this column to move out and scout the terrain where we had just fought.

Generals [Ménard], [Puthod], and von Clossmann had been blocked in their lodgings with all their staffs; however, this was not on the side where first attack occurred. General [unnamed and unidentifiable] informed me, without my asking him, that they had hidden in the cellar. The guard of 12 grenadiers, which I had given them, had fought with courage and prevented the capture of the house in which they were hidden. Lieutenant Hoppen, who General [Ménard][18] had wished to keep by him for several days, and Sergeant Casse had directed the defense. [Ménard's] guard had lost 3-4 men to the enemy's fire. Sergeant Casse was promoted to the rank of an officer, even though he was illiterate, as compensation for the service he had rendered at the quarters of the two generals. It would have been justice to promote Hoppen to the rank of captain, as he had merited it for a long time. I had counseled Hoppen to remain near General [Ménard] so that he could obtain a promotion on the first occasion, but this officer, a Lithuanian gentleman, was as frank as he was brave. He gave the details of the cowardice of the two generals, and it was advantageous to [Ménard] to return him to send him back to his company. I promoted him to adjutant-major of a battalion, in order that he might more easily find his well-deserved promotion to captain.

[18]Editor: The text does not identify this general, but, based on earlier comments, it is Ménard..

The information that I gathered from the prisoners of war confirmed to me that a Prussian column, coming from Danzig, had joined von Schill's troops and that the major had, himself, directed the attack on Stargard. According to the portrait this painted for me of the partisan commander, I cannot doubt that von Schill had spoken to the prisoner to whom I had spoken in my lodging. The attack had been designed to destroy part of the Legion and to take the other part prisoner. He placed, according to the information I gathered, later, from wounded officers that I had left at Stargard, Prussian troops in the farms and in the houses that formed a separate quarter, where the magistrates had not given lodging to the French troops.

A party of Prussian infantry, with the squadron of which I spoke, had moved from the city, to attack the grand guards frontally, while another party of Prussian infantry attacked the left guard from their rear, by capturing the street where General [Ménard] was lodged, in order to open passage for the squadron to enter the fortress, as I have said earlier.

It is incontestable that the Prussian major had hidden soldiers in the houses along the side of the fortress, and that it was these soldiers who fired all the musket shots of which I have spoken, at the same time as the squadron fired its pistols against the fortress.

However, without doubting this surprise, I held four battalions with their backs to the buildings. The battalions, that were concealed in the darkness, saw their officers and commanders walking in front of them to hold them in their ranks and they could not be terrified. As a result, the appearance of the squadron did not put disorder into my men. It then departed as I returned. I cannot doubt that Major von Schill, in lodging all the officers [of the Legion] in a few specific houses, had the intention of taking them prisoner. The night was dark and, if the Prussian squadron had been able to slip behind

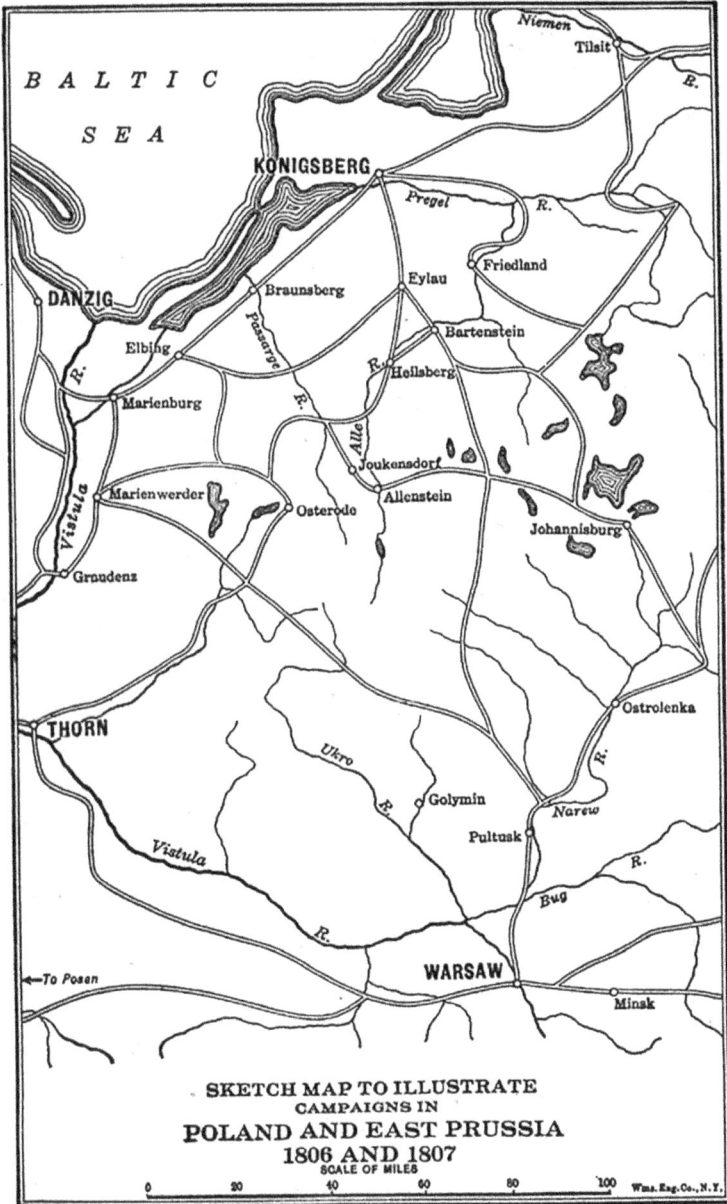

SKETCH MAP TO ILLUSTRATE
CAMPAIGNS IN
POLAND AND EAST PRUSSIA
1806 AND 1807
SCALE OF MILES

the 4th Battalion, which was closest to it would have been destroyed by a quick charge; the three others would have followed this movement.

It is unfortunate, for troops to be under the orders of a general who does not possess its confidence; above all, when the general knows it. The officers engaged me to make a specific report to the major general; but, on one part, this would have been contrary to the principles of military hierarchy, and on the other part, to explain the motives that had caused me to step outside ordinary regulations, I should have been obliged to say that no general had appeared in this affair, and I could only explain it to him or to give a report, which would be the equivalent of a denunciation. I did not know that General [Ménard], who filled the functions of division commander, and the French generosity, which I had on my part, drove from me the idea of implicating the Baden general in an equal report. I could not say what was the force of the enemy, since the affair occurred from 11:00 p.m., to around 2:00 a.m. But it required a large force to attack around the city and inside it. The prisoners of war were in the uniform of Prussian Jägers and that, alone, proved that the enemy troops, which had taken part in this attack, were part of the army corps, which maneuvered before Colburg and Danzig.

The most remarkable affair occurred following the surprise of Stargard, the attack on the city of Dirschau-on-Vistula.[19] The Legion found itself united with newly raised Polish troops under the command of General Dąbrowski.[20] My first battalion was positioned to attack the north of the city. The 2nd and 3rd Battalions remained in line to the west and the 4th Battalion was positioned to the southwest, to

---

[19] 63rd Bulletin of the Grand Army, Ostenden, 28 February 1807. "... General Dąbrowski marched against the garrison of Danzig. He encountered them at Dirschau, threw them back, and captured 600 prisoners, captured 7 cannon, and pursued them closely for many leagues. He was wounded by a musket ball."
[20] This general's name is often spelled "Dombrowski" in English.

stop the arrival of reinforcements thought to be coming from that side. This battalion had an engagement in which it captured a cannon on the battlefield. They encountered Prussian troops, coming from the direction of Danzig, which came to support Dirschau.

The general detached my 2nd Battalion to reconnoiter them. A small affair occurred at this point, while General Dąbrowski bombarded Dirschau.[21] This city's fortifications consisted only of a wall and lacked gates, to judge from the northern side. The affair of the 2nd Battalion became more serious when it was driven back by the first enemy troops to a point behind a undulation of terrain, where they found the main body of their column. I shall enter a few details, in speaking of this affair, because the means by which the enemy was defeated there connected perfectly with the type of tactics that agreed greatly to the French.

I did not see the 2nd Battalion again, and I did not know the abilities of Roummet, who commanded it, which gave me great concern. I asked permission to go see what happened myself, and to lead the 3rd Battalion from this side, but they did not wish to give me but a part of it, with which I joined the 2nd Battalion. This last battalion was in line on the backside of a hill, which had first hidden it from my sight. In front of it, and below the height, was a stream, and on the other side, there was a second, less high hill, on the crest of which stood the village of Milbantz. In front of this village, on our side and on the peak, was a body of infantry, which I judge to be more than 3,000 men, with four cannons, formed in line and which fired shot against my 2nd Battalion.

I ordered Commandant Roumette to withdraw his troops towards the portion of the 3rd Battalion that I was bringing forward, and I sent de Vaivre, at the gallop, to the generals, to request cavalry support. While Roumette made his movement, I placed the four companies of the

---

[21] This is also discussed in, "Memoirs of a Polish Lancer" by Dezydery Chlapowski

3rd Battalion such that they appeared like the heads of four columns to the enemy. Soon, a squadron of Baden hussars arrived, and I placed it behind the right of my line. A small Polish lancer regiment then arrived, under the command of Colonel Franceski, to whom I offered command of the engagement, but he refused it. I could maneuver my troops by column, or deploy them, because my troop knew nothing and Roumette was probably the only officer who knew something about maneuvers. I brought together the officers of the 2nd Battalion to inform them that I intended to deploy the lancers in skirmish order in front of the enemy line, which appeared to be patiently waiting for us. I directed them to explain to their troops that, when a soldier was going to fire, he should move forward twenty paces and then to get between two furrows to reload his musket, fire, and continue to advance in the same manner. After this, upon a musket shot, which I had indicated as the signal for movement, the companies scattered as they ran forward. Their fire was heavy and the skirmishers continuously advanced. After a half-hour I saw, by the clearing of the smoke, that the enemy was maneuvering by platoon. The cavalry wished to charge, but I opposed it. I advanced the mounted troops, with four companies of the 3rd Battalion. This movement fired the audacity of the skirmishers, who threw themselves against the enemy. The enemy withdrew, in disorder, to return to the village, abandoning its cannons, which the skirmishers captured.

It was Surgeon Belhomme who managed the capture of the four cannon. General ...... had initially sent his aide-de-camp ...... with the 2nd Battalion, to have the occasion to bring to him a report, personally witnessed, without a doubt, that he had searched all the cities and the villages where the Legion was going to pass, which I have already said, but he concealed himself, during the entire affair, in a ravine formed by the waters of a stream. I did not have need of him. I did not mention his actions, and, to my great surprise, he was given a decoration shortly afterwards, as if he had taken the cannons himself. Roumette and I were not decorated. We made no complaint and we contented ourselves with the opinion the Legion had of us.

I will now finish the article, which concerns the small engagement at Milbantz.

This village was surrounded by a ditch, which left only a single entrance where the Prussians had retired. The skirmishers followed them very closely and, when they had crossed the village, they scattered, again, to harass them. However, night approached. I stopped the troops and had them take up position, believing that it was too late to take a second village before the enemy could reorganize himself in line to cover his retreat. It was necessary, then, before everything, to know what was happening in the attack on Dirschau (Tczew). I had always thought that it was not absolutely necessary to know the maneuvers prescribed by our modern regulations to defeat German troops, and this small affair, which I have explained, provides evidence that proves that.

There were only a major and 600 soldiers in Dirschau. It seemed to me that, instead of taking this small city, with a strong force, to pillage it and to leave the inhabitants absolutely naked, except for the warning, it would suffice to blockade it, with two battalions, to give it the time to surrender and save itself with a capitulation that would preserve the honor of the major and his troops, and thus save the resources that would be found in the city and for which we had great need. No one knew how General Dąbrowski, and his son were wounded in this affair. Some time later we found a newspaper in which we found a report of the engagement. The emperor gave, it said, 17 crosses [of the Legion of Honor] for the capture of Dirschau and the Legion did not receive a single one, even though it had played a principal role in the capture.

General Dąbrowski did not like the Legion because, on the one hand, he had not wanted that there be French officers in it, and on the other hand, almost all the Polish officers that were in it had served with Dombrowski in Italy and spoke ill of him. Since the capture of Dirschau I had always had contempt for Dombrowski. One must consider him more of a brigand than a soldier. He had been informed, after the capture of Dirschau, that his soldiers had thrown themselves into the city and that the pillage was to the point where the women were chased from their houses, running naked and without shirts in the streets. His response was, "It is good that these children amused themselves a bit." This general took the carriage belonging to Dupont, Surgeon Major of the Legion, to save himself and his son, under the pretext that they were both wounded and obliged to go to the rear to seek medical assistance.

No solider of the Legion entered Dirschau.[22] I had given orders to the officers to hold themselves on the defense. My soldiers all came from the Prussian Army and the troops under Dombrowski were formed from serfs, who had been raised on the lands of the Polish nobility, and who were uncivilized. I know well that similar troops looked to pillage as compensation for their services and that it was frequently necessary to allow them to start it; above all, when they are new, but one could, at least, prevent them from violating the girls and women, and from setting fire to the houses in which they found nothing to steal.

The Polish serfs were lodged in wooden barracks that were more badly constructed than those of soldiers on campaign and then they were led to war believing that all the countries they crossed were organized like the Polish fields; that is to say, that there were only masters and serfs. Then, the natural jealousy which arose because

---

[22] This differs from Chaplowski's account.

of the inequality of conditions caused them to do the worst they could against those who, by their lodging and clothing, appeared to be masters. A Polish soldier set fire to a bourgeoisie's home after they had pillaged it, but he respected the cabin by its side. The opinion of the Poles is that this was a war in which they can avenge the misery in which their masters held them and the blows that they were given to make them work. Such is the true reason for which, all the time, the Polish levy of troops, totally formed from serfs would have gone to the cities to arm and dress themselves, if their masters did not take extreme measures to furnish them.

I was later told, regarding the sack of Dirschau, that, after the entrance of the French into Poland, Dąbrowski had intrigued to have himself raised on a shield, as a Grand Hetman[23] of his country, imitating Jean Sobieski, whose crest is still placed on the charts of the skies, as a constellation. He justified this, since the partitioning of Poland in 1793, because he, alone, had maintain the name, "Polish," in Europe, by means of the troops that he led in the service of the French Republic. Polish songs had been composed that had singularly popularized Dąbrowski's name[24] and this general thought he had acquired the right to become the ruler of his country. However, Napoleon was far from making war to raise this mediocre Polish general and, if this general had great talents, this would have been another reason to keep him idle, instead of allowing him to be raised up on a shield.

In consequence, Napoleon directed humiliating reprimand towards Dąbrowski, and it was to efface the impression that these reprimands were made on the Polish spirit that Dąbrowski had attacked and captured Dirschau as I have said, because the Polish did not know if this city was weak or strong, and Dąbrowski could, as a result,

---

[23] The Grand Hetman was the leader of the country's armed forces.
[24] Dąbrowski's March would become the Polish National anthem

have an exaggerated report published in the newspapers as I have mentioned, to maintain the belief that he was truly a great man of war. The major who commanded this unfortunate city had proposed to surrender within 24 hours, if he did not receive any assistance, and, as he was closed on one side by the Vistula, and on the other by an entire division of troops, his surrender was certain, especially as I had stopped the single battalion coming to his aid. Such were the reasons that General Dąbrowski was not employed in the active army of Poland. He could assume that Napoleon had decided that, not wanting to ruin the country in which he made the war, and wanting, even less, to push the inhabitants into revolt, he could not use such an executioner. However, as he is such an individual, the most ardent and stubborn executioner that had appeared in Europe for many centuries, I believed that the true motive for keeping this general out of Poland was the fear that he would increase his popularity.

Between the capture of Dirschau and the blockade of Danzig there were several engagements of which I shall not speak. I will only mention one of these engagements to give justice to General [Ménard], which he deserves in this affair. I have had so many bad things to say about him that I cannot forget the single action where I found his actions worthy of praise, but this was not a major action. General [Ménard], who commanded the division, wanted to send troops to capture a village, from which the enemy was firing on us, and which was defended by a small river. General [Ménard] objected that the troops were fatigued; that we could take the village when we wanted it; that attacking it would simply result in many unnecessary casualties; that it was only necessary to show troops in position ready to attack it; that the enemy evacuated the village during the night, and that we crossed the river without losing a single man. This counsel prevailed and, the following day, at daybreak, we entered the village without firing a single shot.

It was after the affairs that I have just related that a report was addressed to the Prince Vice-Constable, where the generals asked that he confirm the positions of the officers in whom I had confided those functions since my departure of Mainz, including a young Livonian, a Russian subject, that had been recruited by Hoppen, from the University of Erlangen. This young man was named Yockmann. He had the ability and I used him to translate documents in foreign languages. Soon afterwards, all of those whose commissions as officers had been approved, received brevets signed by the Prince Vice-Constable.

On 8 or 9 March the city of Danzig was invested. The Legion arrived at 10:00 p.m., at the small village of Spikendorf, and, by a large ravine, in another village named Langfurt and which was separated from the city by a promenade, the battalions moved up to the promenade, where we judged we were too close to Danzig. The night was dark, it snowed, and we could neither judge the distance, nor reconnoiter the position. General [Puthod] asked me about what we would do, and we convinced him to leave the companies of grenadiers at the edge of Langfurt, at the entrance to the promenade, while the battalions placed themselves in front of the ravine, so as to bivouac there. The general retired to Spikendorf to take his repose, and I remained with the Legion.

The departure of General [Puthod] appeared singular to me because we considered ourselves the closest troops to the city. The greatest number of the inhabitants of Langfurt, upon our arrival in the village, had fled to the city, because they feared being mistreated by troops whose language they did not know. These fugitives informed the Governor of Danzig that a large number of troops had arrived and of the position they had assumed. The governor knew that we did not know the terrain we were occupying and, still less the formation of the land, and, following the rules of siege warfare, he prepared two sorties during the night, by the Oliva Gate and the Vexelmund Fort, which were united against us before daybreak, and which we easily drove back as they attacked us behind the village of Langfort by the road to the Oliva Abbey.

A sortie by the enemy had been my first thought, but I did not speak of it. I had examined the ravine by which we had descended to Langfurt and I proposed to dig an entrenchment as a reserve position, in case of the need of a retreat. I thought that if a sortie occurred, and if it succeeded, the general would be certain to blame me, and that he would take the affair to his account, if I had been driven back; because he would have time to mount his horse and to join us; at least it is the opinion that I had, since, from his conduct during the siege.

The following morning, we acknowledged that it was impossible to do better in the day. The four companies of grenadiers remained in the position where they had spent the night and entrenched themselves under the city's cannon. The 1st Battalion was placed on the right and barricaded itself there. The 2nd Battalion was placed to their left, behind the houses that masked it. The 3rd Battalion was placed further to the left, vis-à-vis the point of Holm Island, which was guarded by the Russians. The 4th Battalion was placed even further to the left, facing the Neufalwasser Fort, with two companies forward to observe the fort more closely.

We were not in accord with the manner of the attack on Danzig and one could not agree if one did not know the fortifications. Until that point, the French had not undertaken a regular siege for many years. The engineering officers only had the theoretical knowledge of how to conduct a siege, which they had acquired through study. We did not have a single siege cannon and, it appeared to me, that we could do nothing more than blockade the fortress while efforts were taken to seize, by main force, the most forward works of the counterapproach. All along, we seriously wanted to begin the siege works while we awaited the arrival of the cannon that we were to obtain from Stettin, more than 80 leagues from Danzig. Marshal Lefebvre arrived to replace General Victor, who had been captured by von Schill's partisans, as I had said earlier, and we began the communications by which one would arrive at a line from which we wanted to begin the first parallel. Some wanted that the true attack be directed against Bichofsberg and others thought it should be directed against Hagesberg, while even others wanted to begin with the capture of the Neufalwasser Fort.

My advice would have been that we begin by taking Neufalwasser, to dominate the river at its mouth. This was because it was necessary to prevent the arrival of reinforcements and munitions, which could be sent by sea. In addition, it was necessary to blockade Holm Island, by which Danzig could communicate with the sea. This coup de main would have been easy during the month of March. I had observed, the part of the fort that was in front of the right of the two advanced companies of my 4th Battalion from very close up. I had seen that it was only defended by a strong palisade, by a swamp or by pools of water to the front and that the swamps or pools had frozen enough for infantry to cross over them. The Legion would have taken this fort in one night, on the condition that it was given, as compensation, a side of beef per company, and 50 bottles of liquor[25]. I would only have to ask for it.

During the siege of Danzig, all the sorties were made by the Oliva Gate and by Neufalwasser, if one ignores a few early sorties made by the Seydlitz Gate. These latter had, as their objective, the counterapproach works that we were attacking during the night. They moved against the Baden troops, who were on our right; however, the

---

[25] Editor: The term used is "eau-de-vie", which refers to any distilled beverage.

1st Battalion of the Legion moved to the side to support them, understanding that these troops were neither battle hardened, nor sufficiently exercised to act against other infantry, such as those that defended Danzig.

As a result, the Legion received all the sorties that the besieged garrison launched, because, as soon as they were executed from the Oliva Gate, another was made at Neufalwasser Fort, to serve as a diversion. Only one of these enemy sorties was successful and here is that story. I had pushed my 1st Battalion forward into the ally or promenade to the level of the second parallel, and this battalion still occupied a redoubt, in a shallow, which was on its left. The enemy secretly sortied considerable forces by the Oliva Gate. They skirted the Vistula in a low spot, without being seen by an advance post standing in a ditch, and in front of a garden. This sortie, executed briskly, fell violently on the 1st Battalion and overran it.

Thirteen officers were killed in a moment, as well as an equal proportion of soldiers. Captain Miralowski of Grenadiers was captured, along with most of his company, by a mass of drunk Prussians, who threw themselves on his troop. Captain Henry was killed at his post, without wishing to leave the redoubt. From that day, the 2nd Battalion assumed the position of the 1st and the 1st Battalion was placed to the right, in an extension of the attack parallel. This check should be attributed to those who had positioned the 1st Battalion, because they committed the fault of not placing its advanced posts in such a manner that they could observe the Oliva Gate. The fault was attributed to General [unidentified], but he blamed his aide-de-camp [unidentified], which made him even more culpable, since he had, himself, given the necessary orders.

It was at this time, around the middle of the siege, that Prince Michel Radziwill and Count Sobolewski arrived at the Legion in the capacity of colonels. These two young men were sent for the following reason: General Zajączek, with whom I corresponded, followed the rules of military hierarchy, since he was always the first commander of the Legion, and he had informed me, as directed by an Imperial decree, that the Legion was to form two regiments. The first regiment was to have, for its colonel, Prince Michel Radziwill and the second regiment was to have Count Sobolewski. These two men assumed that the reorganization had occurred and came to occupy their posts. General [unidentified] was not happy to see my correspondence with General Zajączek, and he had made frequent reprimands. There is no doubt he wanted to replace them in their senior commands. I could only assure myself that the generals who directed the siege of Danzig did not oppose the reorganization of the Legion into two regiments, but I did not receive any orders on this new formation. It was necessary to consider, on this subject, that one should always be content with the services that this new corps rendered, and that one should fear, in dividing it into two parts, that would then be commanded by two young men with no prior service, it would be difficult to maintain this force in the important positions that it occupied in the siege of Danzig. Until that point, I had commanded the four battalions without the intervention of any chief, but if one had me serve as a major in only two of these battalions, and under the orders of a beardless young man, I should fear a reduction in position, since I had been appointed the major of four battalions, without a colonel, and not of two. This is what I said to General Zajączek.

Until that point, I had not known if the Emperor was happy with the services I had rendered, and if I would still remain at the head of the Legion of the North. I would not have doubted this, if I had not been disgraced before and this sovereign had difficulty giving up his prejudices. He had long forgotten the volunteers, whom the émigrés fought, and who had raised him to his throne. However, it was, with difficulty, that he forgot the friends of the two generals who threatened his throne at the beginning of his elevation, using no weapon other that their reputation.[26]

The Legion had been badly received everywhere it went in Germany, and, in the siege army, it occupied the most perilous posts. From my perspective, I always had in my heart my separation from the 100th Regiment. It was not possible to conceal from me the motive, which I was not told, and I no longer hoped for any advancement. I did not request the Legion of Honor, which was continually given to the soldiers of the corps that I commanded. I despised General [Puthod?] too much to ask for it.

I had the occasion, after the Peace of Tilsit, to see a newspaper, in Warsaw, in which I read that Prince Radziwill had been named Colonel of the Legion. This young man, aged 21 years, had represented that, though he had no military experience, he could command a regiment and, on that assurance, the Emperor had responded, "Be calm in this regard, you will find there a major from whom you can learn military service."

As soon as Prince Radziwill arrived, I gave an order in writing, to recognize him and I had this order translated into Polish so that it could be read to the troops, as the siege service did not permit them to assemble. I considered the arrival of this prince as being proper to inspire confidence in the soldiers, and I intended to use the high consideration that the name of this young lord produced to consolidate the discipline [of the Legion]. General [unidentified] called for me and asked, gravely, if I had given the order to receive Prince Radziwill as colonel. I responded to him that it was His Majesty the Emperor and King [who had

[26] Coqueugnoit remained at Martigues before being named commander of the Legion of the North. He attributed his disgrace to his loudly expressed admiration of Generals Pichegru and Moreau.

Editor: Both Pichegru and Moreau had both aggressively acted to take over the throne of France, and Napoleon saw them as competitors, who jeopardized his hold on the throne.

done so]; that no commander of a corps had ever needed the intervention of a general to recognize his commander; that it was good to recognize him or not to recognize him; that if he wanted to take this latter position, he could give me an order to do so, and that I should destroy the order that I had given. As for the rest, I was the only one who had himself to pity, because of the arrival of the prince as Colonel in the Legion, since, while recruiting it, while organizing it and while disciplining it, I would not have believed that they would be commanded by another, but I was accustomed to injustices, for many years, and I added this one to the others. General [unidentified], to escape the

The Siege of Danzig (Gdansk)

embarrassment in which he had placed himself, as a result of the explanation I had given him, directed, for a second time, the recognition of Prince Radziwill by the battalions and detachments which were not under fire. I then remarked to him, that, according to the law and military custom, he should order me to receive the prince, rather than holding this ceremony himself, and that I should return to my functions. I had my reasons to recognize this Polish lord as colonel, by a written order, and the arrival of Count Sobolewski, who appeared several days later, proved to me that I had done right.

Prince Michel Radziwill was a handsome man, having received a higher education in the Russian court, where he had been chamberlain for three years.

I shall return to Prince Radziwill. His father had held him, for several years, in the Göttingen University, to force him to take more education than was normal for men of his social rank. He had judgment and was capable of learning the military profession. He had never thought of becoming a soldier, because his family was too eminent in Poland to be in the military in any capacity other than as the head of its own troops, and he had accepted service in the Legion, despite his father. This young lord had been very well counseled before arriving at the Legion. He suggested to me that he lodge with me and that he be given two or three hours per day of instruction in the military profession. I accepted this offer, and it was with great pleasure that I gave him the instruction, which was not ordinarily given to Polish officers. However, he allowed himself to be drawn to the headquarters of General [unidentified], where games and the young ladies were more attractive than lessons to a 21-year old man. The general engaged him to eat at his table. The general had a cuirassier, a hussar, and two or three other bad subjects which he sent into the surrounding villages to pillage for food and the prince found the food of the general better than the rations that I provided. From then on, Prince Radziwill no longer occupied himself with his military profession, as if he was not a soldier. Nonetheless, he was very brave. He exposed himself too much, and uselessly, in the beginning of his service. I reproached him frequently, to which he paid no attention. Finally, after I had shown him that General [unidentified] was turning him into an aide-de-camp, and, above all, when I proved to him, in the presence of General Sobolenski, that he sought to either wound or kill himself, he ceased to tell me that the whistling of bullets sounded to him like the buzzing of flies. From his arrival, Prince Radziwill served in the trenches, concurrently with Colonel Larisch, a Baden officer, and myself. I had served in twenty-one siege trenches, for my part.

Count Sobolewski, a very brave and amiable young man, had received a distinguished education. He became bored with being a colonel without a command and without functions, and he returned to the Polish Army, where he was given a regiment, in consideration of the military services his father had rendered under the previous king.

The battalions of the Legion did not move into the trench, as a guard, until early in the siege. It soon came about that the French regiments took this service, as it was preferable to that of a worker. These regiments were placed to the right of the Legion and behind the attack parallel. I must say why the Legion ceased to furnish the trench guards. The 4th Battalion could only furnish them with the greatest inconvenience, because it was placed so as to defend against sorties from Neufalwasser. New works were constructed before the 3rd Battalion, and, independently of that which it was already charged with guarding, it was also obliged to furnish many workers. The 2nd and 1st Battalions were continually involved in the prolongation of the attack parallels, and under the fire of the city's cannon or those from Holm Island.

On one side, the engineering officers tormented, without cease, the generals to have the soldiers from the Legion as workers [to dig the trenches], because they did more work than the French, and that neither cannon balls nor bullets disturbed them. It was disagreeable for a French soldier to be wounded with a pick or a shovel in his hands, but the Polish soldiers did not have the same attitude. They were big eaters and the 15 sols per day for the work given to them was particularly agreeable, since it allowed them to buy additional food.

The French regiments had, over the Legion, the advantage of being peaceful in their camp and being able to repose there, since sorties and cannon balls could not reach them. However, the battalions of the Legion of the North rarely passed the night in their barracks without suffering some catastrophe.

We did not have sufficient gunners to besiege Danzig. I discovered 123 of them in the ranks of the Legion. They had come from the Prussian artillery. I placed Lieutenant Hann, of the Legion, at their head. Hann was a Polish officer, who had served in France in the 1st Artillery Regiment, in the capacity of a sergeant. These 123 gunners were placed in the second parallel, between the batteries of French gunners. Lieutenant Hann directed them in the construction of a battery, which performed the same duties as the other batteries during the siege. The gunners of the Legion could not be commanded by a French artillery officer, because the language problem. Lieutenant Hann won a decoration in the battery, as well as three or four of the gunners. I say, "his battery", because the siege batteries took the name of the officer who commanded them.

The service of the artillery was so tiring, that the officers and gunners alternatively changed between the batteries and the camp every 24 hours, and there were some that were obligated to do double duty in the trench. In this siege, it was obvious that it was necessary to have gunners that were strong, robust, and courageous. I have studied some secondary sieges during the wars of the Republic, but I had never seen the major works that were executed before Danzig.

I shall make, here, mention of a use that I had not known before the siege of Danzig. When the siege was finished, Lieutenant Hann presented me with a unit return that had the name "Nominative State of the Gunners of the Legion of the North, who had constructed Battery Hann, who performed service throughout the siege, and who have the right to a portion of the bells of the city." Lieutenant Hann asked me to certify this unit return so that it might be presented to the artillery commander of the siege. I feared that this was a bad joke and I sent to General Lariboisière a note, asking if the partitioning in question should be made. His response was affirmative, and I later learned, from the artillery officers, the origin of this practice. Since this time, I have read that this was a historical practice; "When capturing a fortress which had required the use of cannon, the inhabitants were obliged to ransom, with money, the bells of the churches and the copper tools and other metal that was to be found in the city; collectively, this was called the bells. All of this metal belonged to the grand master of artillery, who retained part of it and surrendered the rest to his lieutenant, who commanded the artillery during the siege, and to the officers who had served him."

Thus, all the bells and kitchen tools of Danzig belonged, first, to the Inspector of Artillery of France since there was no grand master. His lieutenant was General Lariboisière, and he was generous enough to cite for me the work, which I came to cite, since the simple gunners also had part in the partitioning.

It remained to learn the origin of this practice, and, in this regard, I found that, when the Turkish Army laid siege, it brought no cannons, only copper pigs on camels (they had no horses in Turkey to work the fields), and the *Topchis* or founders [used this copper to] cast cannon of various sizes. The *Topchi-bachi,* or grand master of artillery, had the right to break all the bells of the Christian churches that fell into his power, so as to be able to use them to make cannons and that the Turks so liked artillery that, despite their ignorance, it was they who invented mortars, under Sultan Mahomet II. It was then, truly, from the Turks that the Christian powers adopted the practice of taking the bells of the cities.

In speaking of the distinguished actions by members of the Legion of the North, I cannot forget the 3rd Company, commanded by Lieutenant Tardivel. This company was placed, nightly, in a house on the edge of the Vistula from the beginning of the siege, towards the point of Holm Island, facing a Russian battery, which was established on the other bank of the river. This company entrenched itself so much, with the roof of the house and dirt, that the battery could not drive them out of it; it remained there until, towards the end of the siege, the island was taken from the Russians.

General of Engineers Kirchener, in recognizing the services rendered by there troops, retained the name, *Tardivel House*, on this post and in his memoir of the siege, which was later published. He also inserted a very honorable note on Lieutenant Tardivel in his account. This post was charged with preventing ships from Neufalwasser from reaching Danzig. The Russian battery could not cannonade Tardivel's company when it moved out to block the passage of ships, without striking the ships that were attempting to pass by. It was only during the night that Tardivel could send men to work with the battalion or to receive bread and meat for his troops. Tardivel was promoted to captain in the post that he occupied, but he did not receive any decoration, though one was requested for him.

This officer did not, in this circumstance, do what all other Legion companies of the Legion would have done, if, in their turn, they had been called to the same post. However, he gave a good example to the besieging army, and, by that alone, he deserved a decoration. He was placed, since that time, into the 58th Line Infantry

Regiment. He was, in 1810, at the depot in Paris, where he blew out his brains.

He was unhappy that his talents and bravery had not been recognized in his new regiment. His comrades didn't believe in the exploits that he recounted; he was lively and quick tempered; he frequently had quarrels; he was ashamed to be relegated in a depot, while his regiment was at war, and he killed himself. This officer was an Indian and lived in Pondicherry. The English had brought him to Europe, after the capture of the city. The misfortunes that he suffered, be they in the wars in India, or on sea, where he was taken prisoner and put into servitude, had made him intrepid.

I have already said that the Legion had driven back and foiled all the sorties which were made against it, except for a single one, the results of which I have already recounted, and I have said that I would not speak of these sorties. Besides, the enemy had launched sorties so frequently that I cannot recall them all. However, I owe fair mention of the one that took place by Neufalwasser, and in which the Regiment of Death, under Count Krokow, was destroyed. I do not report this affair because General Savary, today the Duke of Rovigo and Minister, was involved. He was surprised at the good conduct of the 4th Battalion of the Legion, which received this sortie, although this battalion didn't do better than normally. He announced, haughtily, to those that were near him, that he was going to report to the Emperor, and I did not doubt that his report would do much to fix the attention of the Emperor on the Legion, and would give him the 36 decorations that he had accorded to it after the siege of Danzig, at the moment when he held his review. It was, with difficulty, that the emperor had seen the Legion in battle, and, after doing so, he granted it two decorations per company, which came to seventy-two. However, after the review, which lasted three hours, it was related to him that no unit had ever received so many decorations after four months of service and that the other corps would be unhappy, so the Emperor reduced the number of decorations to one per company.

As the destruction of the *Death's Chasseurs* brought much honor to the 4th Battalion, I will explain this action in greater detail. We were occupied in repulsing a sortie by the Oliva Gate, when a second sortie was executed from Neufalwasser, under the fire of cannons and musketry that illuminated that point. I had left the sortie from the Oliva Gate and moved, at a gallop, towards the 3rd Battalion. I had it march out and then went to the 4th Battalion, some of which had already moved forward, towards the grand guards. I then saw the *Death's Chasseurs* before them and I told all the companies that they would have the opportunity to avenge themselves on the commander of the *Death's Chasseurs*, who had executed the soldiers of the Legion that had fallen into his hands.

Three companies of the 4th Battalion, a squadron of the 23rd Chasseur à cheval Regiment, and a few companies of Polish lancers were placed on the edge of the sea, to the left of the pinewoods, and the five other companies of the same battalion, under the orders of Captain Majewski, had moved to the right of the Saspersee Swamp, which was then almost dry (it was 16 April); the grenadier company was detached. Chef d'escadron Lebrun knew that, since he saw the musketry fire illuminating the rear of the sortie, he could have the cavalry charge.

The Lord of Bohn, captain of the 8th Company, 4th Battalion, who came from Austrian service and was newly arrived with the Legion, thought that, according to the custom of good soldiers, he was obliged to make proof of his bravery, in order to establish in the French Army, the reputation that he had merited in the Austrian Army. He threw himself forward with his company, between the fort and the sortie, to cut off the retreat of the Count of Krokow. The other companies were animated by this wonderful example. The cavalry charged on its own and the sortie was crushed by these audacious movements. Other enemy troops sortied from the fort, but the 3rd Battalion of the Legion received them and the Count of Krokow and his regiment remained enveloped. Captain de Bohn lost about 30 men, but this was not too much for such a reckless coup de main.

The Polish lancers wanted to impale the Count of Krokaw on a lance because of the ravages he and his troops had done on the lands of many Polish lords, but a maréchal de logis [sergeant] of the 23rd Chasseur à cheval Regiment stated that French soldiers predominated in the siege forces and that French practices, therefore, prevailed, and that by this reason, his troops would not suffer to have a prisoner of war spitted on a lance, in the manner of the Turks. The Count de Krokow had already suffered a saber blow to his cheek and several other wounds. He was led to headquarters with several other officers, who were also covered with blood and were drunk. One of them insulted the generals in his fury. The Count of Krokow was exchanged about two months later. General Drouet came while the troops were engaged by the enemy. He found the companies of the 4th Battalion, but I remarked to him the cannons of that point of the fort did not dare to fire at them in fear of hitting their own troops and that it was for this reason that the famous Krokow had sortied with artillery. The artillery, the horses, which drew it, and this commander were led, with all that remained of his troops, [into captivity].

At the end of this affair, a number of generals and staff officers arrived, as was normal, and each pretended that he had done something. It was, in this affair, what I saw as proof that the Legion hated General [unidentified] and his aide-de-camp [unidentified]. Thirty skirmishers of the 3rd Battalion, more or less, did not fire, because the enemy returned into the fort and was already too far away. General [unidentified] sought, after a while, the opportunity to decorate his aide-de-camp [unidentified]. He sent his aide to these skirmishers so that they might go back by a movement on their right, and this aide-de-camp was shot in the arm during this movement. The soldiers thought that he was the individual who passed through the villages assuring their bad lodgings, as I have mentioned earlier. This wounded aide-de-camp did not receive any decoration because several generals had seen that the mission, with which he was charged, was not in the least dangerous.

It is for the Legion, and not for myself, that I write this historical memoir, since I intend to have it written by one my secretaries, who has a good hand, and to send it to the colonel who now commands the 5th Infantry Regiment of the Army of the Grand Duchy of Warsaw. However, I believe I must report here a circumstance, which proves that I had the confidence of the troops.

Towards the end of the siege of Danzig, a Russian division of 10,000 or 12,000 men had disembarked at Neufalwasser to move to the support of the city. We did not know if this division would attempt a passage by Neufalwasser, or if it would make its efforts on the other side of the Vistula, by the Vexelmund Fort and Heubuden Island. These two forts are separated from each other only by the Vistula. Oudinot's grenadiers had arrived to assist in receiving the sortie and they were held near Neufalwasser, because it was on this side that the Russian division could most easily make a hole [in our lines] to penetrate to Danzig.

Gardanne's division, which occupied Heubuden Island, was too weak to resist a Russian attack, and the only French troops it contained was the 2nd Légère Infantry Regiment. As a result, it was decided to send it reinforcements. The 3rd and 4th Legion Battalions and six cannon were designated to form this reinforcement. It was necessary to cross the Vistula during the night, in order not to be seen by the cavalry at the Oliva Gate and by the other batteries, which could impede the passage. This passage was even more difficult, as there were only three fishing boats available to move the men over. In addition, there was a raft for the transport of six cannons, six caissons, and the horses.

At that time, Général de brigade Jarry[27] was in the army's headquarters, having been sent to repair an error committed before the Emperor. This fault was not having overcome an obstacle by deploying the battalions of his brigade, and the Emperor had removed him from command. This error was not very remarkable, but General Jarry came from the Army of the Rhine[28], and the Emperor was very pleased to do this, or, even if it was not true, as one had always said, that the generals of this old army were better instructed than those of the others. The Emperor has often said: "Only Schauenbourg, of this army, knew how to maneuver!" General Jarry was advised to ask to be employed in the siege in Danzig, instead of asking for a pardon, and his request was accorded. The Emperor had judged that he could not continue in the major operations of the campaign before the capture of Danzig, and that everything that accelerated the reduction of this fortress was agreeable to him. Colberg, Danzig, and Graudentz were, in effect, three points of support on which one could maneuver an army that disembarked into the rear of the Emperor, if it were landed in Old Prussia before

---

[27] Étienne-Anatole-Gédéon, Baron Jarry, Général de brigade on 21 February 1807.
[28] Editor: In fact, Napoleon had a personal reason to dislike many officers from the Army of the Rhine. This army was commanded by his two major rivals, Moreau and Pichegru. Many officers of that army were ardent supporters of both men and they suffered, in their careers, because Napoleon thought they retained a loyalty to either man. Major Coqueugniot speaks of this directly in regard to Elbé earlier in this manuscript. See page 8. On page 35, Coqueugniot speaks highly of Pichegru, which confirms the editor's belief that he was a partisan of that faction and his career suffered because of that.

the capture of Danzig.

I now return to General Jarry. The disembarkation of the Russian division promised a difficult fight. We wanted to profit from this occasion to write the name of General Jarry into the report that would be written, and he had been placed at the head of the Legion's two battalions, which were sent to reinforce Gardanne's division.

Panoramic view of the Siege of Gdańsk by French forces in 1807
by Jean-Antoine-Simeon Fort

Towards the evening, the two battalions were sent to the banks of the Vistula and, when they saw that they were to be taken to the island, under the orders of a general that they did not know, they asked for me in Polish. I was in the trench that day. I was sent an order to put myself at the head of these two battalions and I brought them across the Vistula, as well as the cannon and their caissons. Since the troops had strongly complained about General Jarry, I counseled him to stay at a distance while the passage was affected, causing him to observe that the habits of the soldiers were such that it was useful that I cross on the last boat, and that we cross the river together. When we arrived on Heubuden Island, General Gardanne ordered that the 4th Battalion be placed at Neufer, towards the Pilau point, because enemy troops were expected there and the 3rd Battalion remained with the main body of Gardanne's division.

The Russians advanced from Heubuden Island and I promptly called back the 4th Battalion to fight with the 3rd Battalion and the 2nd Légère Infantry Regiment. The attack was completely repulsed. The journal of the siege mentions this affair. It was brilliant for the French troops. We counted around 3,700 or 3,800 Russians on the field where the battle occurred. General Gardanne ordered me to receive the division's muster and our losses were 518 men missing from the muster.

This disparity in losses was shocking. The Russians had presented themselves in battle, packed like herrings. The 2nd Légère Infantry Regiment and my two battalions threw themselves out, in skirmish order, against the Russians. The Russians could only fire uncertainly, as their own smoke concealed their adversaries from them. Our field guns fired canister into them at short range. A pine woods, which was in our lines, masked our movements. General Schramm[29] performed prodigies of valor and the 2nd Légère Infantry Regiment had a blind confidence in him. Brayer, colonel of this regiment, surpassed himself that day. Madame Schramm, dressed as an Amazon, galloped through the middle of the shot, shell, and smoke, under the pretext that she was there to tend to her husband in case he was wounded (I could not get Madame Schramm to return to the pine woods.)

This affair was, as a consequence, like that of Milbantz, of which I have already spoken. There was

---

[29] Jean-Adam, Baron Schramm, Général de brigade on 24 December 1805, Lieutenant général on 10 March 1815.

nothing shocking about the Russians being unable to stand before such thunderbolts and that they had lost 3,000 to 4,000 men. Oudinot appeared towards the end of the affair, at the head of a battalion, which held the right of his converged grenadiers and voltigeurs. The passage of the Vistula had prevented him from arriving sooner. He removed Gardanne from command, telling him, "I am here." Oudinot's horse had its neck pierced by a bullet and he immediately mounted another horse.

It was pointless that Prussian maneuvers had been introduced into the French Army. The tactics by which the country of Hugues Capet [France] had enlarged the kingdom before the Revolution were always better for the French. Pichegru appears to me to have been the only French general, in more than 100 years, who was persuaded of the truth of this.

Generals Oudinot, Gardanne, and Schramm, as well as Colonel Brayer were majestic in the affair that I have reported. The composition of their figures, their movements, their words, and everything else, with brave men, gave a thunderous implosion to their troops. I believe that such a brilliant soldier, whose manners were noble in an affair, and who carries his head so majestically, can only be so as a result of his habits in battle, joined with a bravery sustained throughout the campaign. One would have to be a coward not to behave so courageously and calmly when one has such an example before their eyes. These examples were much more striking for me, as I was in the habit of seeing daily, on the other bank of the Vistula, the behavior of certain generals and commanders, in front of the enemy, that froze the heart of their soldiers.

I was on Heubuden Island and there was no one with the 1st and 2nd Battalions who could direct them to implement their orders. I received a written order to return, but General Gardanne, in whom I had inspired confidence, said that I was under his orders with the 3rd and 4th Battalions, and opposed my departure. Finally, a letter came from the chief of staff, directing me to recross the Vistula. Prince Radziwill did nothing for the service; he didn't care to, either, and the 1st and 2nd Battalions found themselves without a commander.

I will now report on an event that was remarkable for its uniqueness. We saw, a few days after the affair I have just recounted, a post with 50 men from the Legion, attacked by an English corvette armed with 24 cannons, and the same corvette stopped by a company of voltigeurs from the 4th Battalion of the Legion, a few minutes after it drove back the 50-man post. This post was placed on the banks of the Vistula and entrenched by a simple elevation of dirt on the waterside. It was commanded by Lieutenant Kaminsky, aged 60 years. I will explain these events in a more detailed manner.

Towards 5:00 p.m., and the same day as I returned from Heubuden Island, the corvette sailed before the wind to Neufalwasser and moved up the Vistula to enter Danzig, firing canister to the starboard and port, to hold the infantry on both banks in their trenches along the Vistula. When this ship arrived in the vicinity of the 50-man post, Lieutenant Kaminsky, instead of drawing his 50 men behind their entrenchments, while it passed, spread them out, in skirmish order to engage the corvette, taking it in the rear, so as to avoid the canister, he then moved to withdraw, and he moved with his troop to the measure that the ship advanced. Commandant Vanrosen of the 1st Battalion, who was near there and who was unhappy seeing our posts flee in front of the ship, detached his company of grenadiers to have them fire on the 50 men if they decided not to return to their post.

While this movement occurred, the corvette arrived within cannon range of Redoubt No. 5, which was commanded by Captain Bellanger, commander of the voltigeurs of the 4th Battalion of the Legion, who formed the garrison of the redoubt. The guns of this redoubt began a sustained fire against the corvette, but it continued moving on Danzig. Soon, it was so badly pounded by the shot and canister that, in order to move away from these cannons, it moved into an elbow in the Vistula, formed by Holm Island, nearly captured by the Russians. The Paris Guard, dressed in red, and which the other troops called, for this reason, the *écrevisses* [crayfish], occupied the point of this island.

The corvette fired on all sides, and one did not see a single soldier's head show itself above the entrenchments. As the sails of the ship were furled (a precaution which the sailors take in various cases), the corvette could not move, as would have been necessary to clear the elbow in the river and to continue on its course. It hoisted the signal that indicated it surrendered, but Captain Bellanger, who did not understand that signal, continued his fire. Finally, the signal was understood, and he ceased fire. The ship found itself on the right bank of the Vistula and near the Paris Guard. These troops came out of their trenches. They climbed onboard the ship, taking the wine, rum, and food that they found there. When Marshals Lannes and Oudinot arrived, the soldiers were already drunk, and it was impossible to get them out of it.

I also went there, and we left promptly, fearing the soldiers would blow-up the ship, because its cargo was gunpowder and munitions of every type. I came back to the ship with Marshals Lannes and Oudinot, the captain of the ship walking between them, and, when we arrived at the bridge of rafts constructed over the Vistula, the sailor stopped, out of surprise. He said to the generals that had he known that there was a bridge, he would not have attempted the passage. Other ships were in Neufalwasser, and they only waited for the arrival of this ship in the proximity of Danzig before they began their own movement. So, the English captain knew better than the

French naval officers that, before Danzig, the weight and speed of the ship would break a bridge of rafts, supported by cables, like that of which I speak, because he spoke and explained his reasons, without hesitating. Meanwhile the officers of the French Navy argued over this important question.

The crew of the corvette consisted of about 250 English sailors, all drunk. There were many wounded. There were also female passengers on the ship, including several who were wounded. I do not recall how many dead there were. Found among the crew was an aide-de-camp to General Kalkreuth, who had come out of Danzig to give, to the fleet that had delivered the Russian division, the telegraphic signals by which the ships could communicate with the city, but no one else could understand it.

The account of the siege, written by General Kirchener, says that "the corvette was forced to lower its colors by the fire of the infantry, which came from the banks of the river." This account is far from the truth.

This account is far from exact. There was not a single musket shot made at this ship and there was no other movement by the infantry except that of the 50-man post, the grenadier company, and the 4th Voltigeur Company, of which we have already spoken. The Vistula is not large between Vexelmund and Danzig, but it is very deep, but it has almost no embankments. The terrain is very plain and the fire of the cannons of the corvette swept it. This was why I did not see a single soldier raise his head above the parapet of the entrenchments, except the two companies and the 50 men of the Legion of whom I have already spoken.

I shall report two more affairs in which the troops of the Legion distinguished themselves in a very remarkable manner. The siege works found themselves very advanced at the beginning of May and, by the 5th of that month, the principal attack was within 6 toises of the salient of a demi-lune towards which it was directed. Then it was judged useful to capture Holm Island and the redoubt close to the Oliva Gate.

The detachment which was to capture of the island was formed of Saxon, Baden and French troops, and it had 300 men from the Legion of the North, commanded by Captain Sprunglin, a Swiss national. The detachment charged with capturing the redoubt was composed in the same manner. The 4th Grenadier Company of the Legion was in this force, and it was commanded by Captain Pallandre. The conduct of the detachment charged with the capture of the redoubt was confided in Colonel Larisch, a Saxon, and that of attacking the island was commanded by Adjudant commandant Aimé. It was during the evening of 6 May that the detachments were formed in front of Langfurt, so that they could set out at midnight the same day.

In the detachment of 300 men from the Legion, there was a young officer named Nowiski, aged 19 or 20 years, and who was gifted with a rare courage. When the detachment penetrated into Holm Island, Nowiski's peloton found itself engaged with the Russians. The night was dark. One could not recognize a soldier until you could touch them, and Nowiski, who was in front of his troops, found himself in the middle of the Russians. He cried to his men, in Polish, "Friends, I am in the middle of the Russians, fire on us!" The Legion soldiers fired. Nowiski was killed and we did not discover, until the morning, if this brave officer had been killed by his soldiers, or if he had been killed by the Russians.

I have read, since, in several works, that this heroic action was by a sergeant of the 2nd Légère Infantry Regiment, but that is false, because the 2nd Légère Infantry Regiment was on Heubuden Island and it took no part in the capture of Holm Island, because it had been assigned to the duty of repulsing any sortie that the Prussians might launch by Vexelmund; 2nd, in the obscurity of the night, a French sergeant would not have known if the soldiers were Baden, Saxon, Poles from the Legion of the North, or Russians; 3rd, Captain Sprunglin (who had one hand pierced in this engagement), had come to find me the day before, in the morning, in the camp of my 4th Battalion, where I had passed the night, and recounted Nowisky's story to me, with all the circumstances; 4th, finally, Nowiski knew Russian, and, even though he would not have been fluent[30], he would have recognized the soldiers among whom he stood, because the Poles and Russians get along.

Without a doubt the sergeant of the 2nd Légère Infantry Regiment was very capable of doing this, as was Nowiski, but it is necessary to recognize each for what they did. Why not accept that a young Polish gentleman is capable of doing the same as the Chevalier d'Assas, a French gentleman who won the honor of an engraving and the homage of historians? D'Assas appears to be to have been a negligent officer, ignorant and careless, since he had allowed the post he commanded to be surprised. In contrast, Nowiski attacked the Russians audaciously, after having crossed the Vistula in a poor fishing boat.

The redoubt was guarded by 368 Prussians. Its ditch was large, deep, and full of water. It was fraised[31] and palisaded. Its gate faced the Vistula, and, in order to get into it, it was necessary to cross the canal, which one can see on Kirchener's general plan of the siege. The canal is large and deep along all its length, and wood for construction was floated down it. The redoubt was armed with about fifteen cannons, I'm told. Colonel Larisch had surrounded it, but the continuous fire of its guns did not permit the assaulting troops to stand up.

---

[30]Editor: I am uncertain of this translation, as I cannot find the term used, "sçue", in any dictionary. It is either a typographical error, or an obscure term in an archaic spelling. The original manuscript has numerous archaic spellings, as well as some typographical variations of terms.

[31] Editor: A fraise is a palisade placed horizontally, points aimed at the enemy.

Martres, Chef de bataillon of the Legion, had gone to see the expedition out of simple curiosity. Pallandre, captain of the grenadier company of the 4th Battalion, proposed that his troops cross the canal on their bellies, floating over on logs, in order to reach the gate of the redoubt. He rode the first of these logs, while crawling toward the other side, and his troops followed him. Martres crossed the canal in this manner and reached the gate of the redoubt at the same time as his company, but, unfortunately, this gate was closed from inside, with a padlock, and the palisades which formed it could not be broken with blows of an ax, because they were too big.

Orlewski, the company's lieutenant, suggested that, if one could lift him over the gate with a hatchet, he could break the padlock and then they could enter into the redoubt. Martres was extremely strong, and he took Orlewski, who held a hatchet in his hand, and threw him into the redoubt. Orlewski broke the padlock and opened the gate. While this was occurring, Crasiski, a simple grenadier and a man of colossal stature and extraordinary force, slipped down the path, armed with an axe. He moved against the fraise and arrived at an embrasure close to the gate. He worked his way into the redoubt, while frightening the gunners with his cries, and, while he struck at them with blows of his axe, the company threw itself into the redoubt by the gate, firing and obliging the Prussian infantry to lay down their arms.

Crasiski had a decoration for his bravery, but this was not until after the transfer of the Legion into the Polish Army. I cannot omit to say that Orlewski was shot twice while he was in the air and falling into the redoubt.

When awards were given to the troops of the Legion who took part in the capture of the redoubt and the island, Captain Sprunglin, his drummer, and another soldier of the detachment of 300 men, received a decoration, but not a single decoration was given to the 4th Grenadier

Company. Shall I say why this company did not receive a single decoration?

[Unidentified], the aide-de-camp to the general, the same who had been wounded by our soldiers, as I have said earlier, was not decorated. His family was known by the commanding general and they sought to insert his name on the report of the action, in order to obtain for him a decoration. As a result, he was counseled to go with the detachment charged with taking the redoubt. Chef de bataillon Martres, Captain Pallandre, and the two other officers of the detachment, told me, 20 times, that this aide-de-camp had not appeared there until a good hour-and-a-half after the reduction of this little fortress, and that no other man of the detachment had seen him before us.

General Crasiski was a very big eater and, to compensate him for having been so poorly compensated for his bravery, I had him issued a double-ration from that date on. This gratification was more agreeable to him than a red cordon and his comrades thought him most fortunate to have been compensated in this manner, because many of them later asked their Captain Pallandre to judge their actions so that they might be equally compensated. One cannot doubt the truth of this act, when the same Grenadier Crasiski, after having received a Polish decoration at Zakrozim, during January 1808, proposed to sell it to his captain, in return for the moderate sum of 3 ducats.

Captain Pallandre instructed me of this proposition. I had the grenadier brought to me, on the pretext that I wanted to buy the cross, to present it to a corporal who had been my assistant, and I asked him many questions on the reasons why he wanted to sell it. His responses were pretty much as I am about to relate. "Such things as this cross are only good for officers, because they have plenty of money, but they are useless for soldiers, because they are not sufficiently well dressed to wear silver crosses attached to ribbons of silk; that the compensations that are most agreeable to soldiers are such things as new coats, new pants, good shoes, stockings and gaiters, white bread, meat, lard, liquor, not wine, and beer, and money given in the morning so that it won't be stolen, etc; that the soldiers will volunteer for actions, if one wanted to give them, every day, the things indicated, but that it is necessary, to above all, give them liquor three or four times a day, and to not force them to get up when they are sleeping."

I asked him if he was not pleased to be distinguished by a cross that his comrades did not have and his response was, "the happiest is he who has the most money; that he preferred his double ration to any cross of the Legion [of honor], and that he wanted only to have money, to buy schnapps every day."

I return to the account of the siege to discuss the assault, which was ordered at the moment when it was thought possible. The Legion, as all the other troops, had furnished detachments, more or less strong, to the formation of three assault columns. I shall report, only, that I entered the last trench and that I commanded it. The commanding marshal had given a 24-hour truce to the Prussian commander, General Kalkreuth, and General Ménard, who was to be commander of this trench, was absent to summon the commandant of the city. He obtained it, in effect. It was, perhaps, necessary that the generals went to see the marshal to know their destinations, or at least to ask that he assign them. With regards to which I shall say about Général de division Michaud, former commanding officer of the Army of the Rhine, was the only man who had camped in the midst of his troops during the siege, in order to better observe operations, and who made no effort to have command. It was no doubt humiliating to him to put himself before Danzig, under the orders of a marshal, who had been nothing more than a trivial and ignorant *adjudant général* at the time when he, Michaud, had been a commanding general.

I should report an unusual incident, which involves the last trench, because it demonstrates a practice that is little known in the infantry. According to this practice, which is certainly an old practice in the artillery, since it is necessary that one cleans the cannons when they cannot be fired; that is to say, they are always held loaded and ready to fire and the same one must discharge them when the besieged fort's surrender is certain. The artillery commander came to me to ask permission to proceed with this cleaning of the guns, which consisted of firing them into the ground.

The fire began soon after, in all the batteries. The Prussian gunners, who knew this practice, did not respond. It was thought, at the French headquarters, that the siege was continuing, or rather, they did not know to whom to attribute the cannonade that they heard, and the aides-de-camp arrived, at the gallop, to ask me why the firing continued. I was blamed by some generals for having permitted the cleaning of the cannons, but those of the artillery supported me.

The 1st Battalion, for and in the name of the Legion, made its honorary entry into Danzig. The columns, which preceded it in this military ceremony, were formed of the battalions or detachments drawn from all the units that had taken part in the siege. The three other battalions of the Legion were posted at Neufalwasser, where they were very badly treated. There was much discontent with the troops because the Baden troops were much better treated. I heard a general accuse General Ménard, overtly, declaring himself the protector of the Baden troops, because he wanted to win the great decoration of the Margrivate and he eventually received it. The crown prince to the Grand Duchy was always at the siege of Danzig, at the head of his father's troops, and he was known to have particular concern for his troops.

There had been taken, with regards to the Baden prince, a political measure, which would not be done for a simple soldier of his grade. From the moment when this Lord appeared at the siege army, an order from the major general was published, which directed the generals to prevent His Highness from entering the trenches and which held them responsible for any accident that might befall him. This young prince was the only son of the Duke. His presence was not necessary in the works, since he neither commanded a brigade, nor a division, and it was in his best interest that he was prevented from exposing himself. Besides the public must believe that the education received by the princes give them a force of spirit and an ardor, which allowed them to brave dangers that commoners feared, and he assumed the political posture to encourage this belief. Sometime after the capture of Danzig, the Emperor consulted the Grand Duke to go impregnate his wife. He had married, shortly before, the Princess Stéphanie, the adoptive daughter of Napoleon.

Two days after the capture of Danzig, the Emperor arrived there, on horse, and was followed by a great number of princes, marshals, and generals. He wanted to see, in the field, the Legion. I immediately left Neufalwasser with the 2nd, 3rd, and 4th Battalions to lead them to Danzig, where the 1st Battalion stood. The Emperor grudgingly allowed me time to place the battalions in columns by *pelotons,*[32] in the street where [the Emperor] was lodged. General [unidentified] took command. Prince Radziwill and I shrugged our shoulders and laughed. The princes and generals, who were on the staircase of His Majesty's lodging, understood the situation and laughed like us, after I cried out, in passing alongside them, "Nothing remains for him to do but to put his sword in its scabbard!" At that point, my voice was the only voice that the battalions knew

---

[32] Editor: A "peloton" is the tactical name given to a company, the company being an "administrative" organization. There was, however, no difference. "Peloton" solely refers to the tactical employment of the administrative company.

as commanding, and the officers, as well as the troops, were shocked that this cowardly general put himself in my place. I decided, at that point, to ask the pity of the major general, but by a delicate sentiment, well or ill-placed, I decided to do nothing. I feared, on one part, that one would only see, in this, that I acted out of envy, and of the other, that it was my intention to point out the uselessness of Prince Radziwill. Without a doubt, the Emperor and his major general well knew that the prince did not command the Legion, but was it not in their view to save him from this humiliation in front of so many generals?

The review lasted less than three hours. General [unidentified][33] spoke continually to the Emperor. But this sovereign did not say a single word to him. When he had questions he had, he addressed them to Prince Radziwill, to whom I gave the answers, when necessary, and this was the manner of acting in a hierarchy. The Emperor avenged us, thus, from the intentions of General [unidentified]. At the end of the review, the Emperor spoke to me for a long time. He asked me on every aspect of the service of the Legion and, principally, on its needs and resources. He asked me why we had not received the 518,00 francs that I told him we were due. I responded that we would have been embarrassed while the Legion was fighting. This response was applauded by some of the marshals. One of them observed, to those that were near him, that he had more than one chief that the 518,000 francs would not have embarrassed [in such circumstances].

The Emperor asked, during the review, that he should be presented with two men chosen from each company, to whom he would grant a decoration, which came to 72 for all of the Legion. However, after the review, the number was reduced to 36, for the reasons that I have already explained. The ribbons were sent some time later and there was not the least anger over the reduction of

---

[33] Editor: This is probably General Puthod, based on the actions and the author's well-expressed attitude towards him.

which I have spoken.

The decorations of the Légion d'honneur, accorded during the siege, were raised to 11. Those presented by the Emperor after the capture of Danzig came to 36. This meant that the Legion won 47 decorations.

Subsequently, the Legion left the service of France to become part of the Polish Army; the military diet of Warsaw accorded it the decorations of the Polish Military Cross as follows: Chevaliers of the cross – 4, Gold cross for officers – 6, and Silver crosses for the NCOs and soldiers – 8; a total of 18, giving a total of 101 decorations won at the siege.

It is well known that I received the Polish decoration of the Chevalier [of the Polish Military Cross]. I was the only senior of the Legion who had been compensated by a title and a majority. I was named Chevalier of the Empire, with an annual rent of 2,000 francs in Westphalia.

On 2 June 1807, the day of the Emperor's review, the clothing of the Legion was in the most pitiable condition, the coats, vests, and their breeches which General Clarke had given me were in rags. The soldiers were black and dirty. The felt of their shakoes was faded white. For five months, the troops had slept in the snow and mud. The work in the trenches had rendered everything dirty, destroying it all. No replacement had been made. The Emperor, in expressing his displeasure, observed to those that surrounded him, that it was wrong to keep such badly dressed troop in the field. He ordered the Legion be newly dressed, drawing the cloth from the Danzig magazines. His orders directed that the dressing [of the Legion] be completed within a month.

The measures taken to execute the Emperor's orders progressed slowly. I encountered obstacles everywhere. I always believed that my efforts were obstructed by influential individuals, who desired that the Legion was not employed in the war that was about to begin on the Vistula, a war that was not stopped to wait for the capture of Danzig. I saw, on all sides, a force of inertia that I could not defeat. The great majority of the officers were discontented. None of the commanders had received any decorations, other than Prince Radziwiłł. One Lord Pallandre, a very handsome man, captain of the grenadiers of the 4th Battalion, had well served the Legion and was designated by me to take charge of the reissue of clothing. He spoke with much confidence. He had taken ascendancy over the officers in a freemason lodge, of which he was the founder, and I believe the general was persuaded to take him into his confidence.

This general had stolen two horses, which a Stettin merchant had furnished, in February, to carry our baggage to Stargard, and it was easy for him to obtain a carriage to Danzig. Two of his servants, eight days after our entry into Danzig, were already dressed in scarlet cloth, stripped with gold. I subsequently knew that it was the officer in charge of matérial[34] of the Legion who had dressed them.

I rode in a carriage to the general's lodging, and one said, every time that I was there, that the Grande Armée had rested, in villages, during the whole time that the siege of Danzig lasted; that the siege troops had completed their task and that the Legion and Baden troops were no longer obliged to continue the campaign beyond the Vistula. All this gave me serious concern, because it seemed to me that all the world, except me, was united to prolong the provision of the clothing [to the Legion], and that, if the Emperor had ordered that the cause of this slowness be uncovered, one would have blamed me, because Prince Radziwill was absolutely hopeless. I had not wanted to be his consul, his editor, and his secretary. All orders were given in my name and signed by me during all the time that I remained with the Legion.

---

[34] Editor: The original term is "*capitaine d'habillement.*" This rank does not exist in English-speaking armies. He is responsible for all the equipment of a regiment, including weapons.

I had attached very intelligent officers to work with Captain Pallandre, and it did not appear to me possible to do better in the prompt provision of the necessary equipment. Every time that I grew angry, or that I gave orders to reprimand the slow workers, someone objected to me that the material was confided to the city's artisans who worked with them, that this was necessary, as some who were so employed did not know how to make uniforms. In these circumstances, I thought that, to cover my responsibility for this, that I should, from time to time, dictate, on the register of the consul, explanatory deliberations on the insurmountable obstacles that the preparation of the uniforms endlessly encountered.

The discontent in the Legion manifested itself when it was necessary to send a battalion to Heubuden Island, towards the Pillau point, where it was said that the Russians had made a landing. The 3rd Battalion was designated to take part in this expedition. I gathered it together and dispatched it, with great difficulty. At this time, I could make no more promises, and the mass of discontented officers had much weight. Should I have believed that Prince Radziwill, heedless as he was of it, had pretended to take up arms, and did it to show his compatriots that his family made common cause with them? Should I, then, believe that this young prince, who had been chamberlain at the Russian court for three years, desired that the Legion be promptly placed in a state to fight the troops of this sovereign to whom his family had the greatest obligations, and under whose domination lay three quarters of his father's fortune? On the other side, what role would he play, if he had been made a prisoner of war by the Russians? His mother had, for a long time, been the first lady of the Russian court. Should I believe that the letters that she wrote him had suddenly excited him to turn an old friend into an enemy?

It was, at the same, time that there arrived, from all sides, many officers who had nominations to enter into the 1st and 2nd Legion. The cadres were soon filled, because the Emperor had confirmed, in their grades, the provisional officers that I had received while crossing Germany, as I have said earlier. We had taken the position to not receive the newly arriving individuals who exceeded the authorized complement, and they were sent to the Vice Constable to be placed elsewhere. However, the prince returned them to the Legion, with the order to place them *à la suite* all those who found themselves in this situation. We were authorized to have 128 officers, but it soon rose to 200.

During the entire siege of Danzig, and after the capture of the city, I held control of the officers of the Legion, and I commanded the service of all the guard tours. The moral state of the troop was as follows: There were 53 officers assigned to all the service tours, because there were 34 killed and wounded during the siege. It is well known that the losses of non-commissioned officers and soldiers was proportionately greater and that, without the successive arrival of a large party of soldiers that I had left in the rear while crossing Germany, the Legion would have entirely melted away during the siege of Danzig.

I gave great importance to personally commanding all the services, in order that the four battalions should equally contribute officers and troops for them.

When I was in the trenches, the orders, which I had given in the parallel or in the zig-zag where I was found, were written down with pencil. I saw several soldiers write their orders in a similar situation.

I always remarked that no one read in the parallels or the zig-zags. One found there the spirit that excluded anything relating to reading. One can [also] believe that women are braver than men, as I have seen vivandières [sutlers] spend day and night with the troops in the trenches, and I have never seen a single man remain there without being obliged to.

The Legion was issued new uniforms when the Governor of Danzig (General Rapp) gave us the order to consult this corps to know if it wished to remain in French service, or if, to the contrary, it wished to enter into the Army of the Duchy of Warsaw. A request of this type necessarily occupied everyone. The French officers did not wish to leave French service. The Polish officers, except for a few who had previous service in the French Army, and who were without fortune, desired that the Legion be passed into French service. Prince Radziwill did not speak to this point, but I was well persuaded that he desired that the Legion request to enter Polish service, because he had appeared to me to have been greatly pleased to have been named chamberlain to the [Russian] Emperor. In speaking on this subject, I said to him, "I am persuaded that, after consulting the officers, all those who are French wish to remain to remain in French service and that all those that are Polish wish to enter Polish service, but the soldiers would do what he [Prince Radziwiłł] wanted, since they were all drawn from serfdom, and had nothing to hope for in Poland, so it mattered little to them if they served the Emperor of France or the King of Saxony." I commented to him that the principal point was to know what the Emperor wanted. I encouraged him to go find General Rapp, to ask him what he thought in this regard. I did not judge it proper to go with him, so it would not be seen that I was the author of this action. He went there and told me that the general had given him a vague response that, in allowing the Legion to vote, one should not act on a minority of the votes, and that it was desirable to recognize the voice of the mass of the soldiers. As this response was not well explained, I presented to the prince that it was reasonable to leave each man the liberty to ask what he wanted to do, and we decided not to influence anyone.

General [unidentified] did not see things like us. He desired that the Legion vote for French service. He continued to hope that he could become commander of it, since it formed a total strength equal to two regiments, by virtue of its numbers and counted as a brigade if he held it under his orders, and he believed the Emperor was suspicious of him as a result of his difficulty in obtaining a command in the line. The Poles discovered his project. They hated him and Prince Radziwill despised him beyond my ability to describe. He hated him so much that, if their patriotism had not caused them to seek Polish service, they would have acted solely to counter the views of this general. Many of the French officers thought like he did since they feared losing their active service and because they desired to assure their position. Some spoke with me of a project they had formed to influence the votes of the 1st and 2nd Battalion, which were at Neufalwasser to vote for French service, but I avoided giving it my support. I responded only that I did not want to recruit for the King of Saxony, nor against those who might desire to determine the Legion to enter his service. I wanted to listen to everyone and to reserve to myself the right to speak frankly, in case of need. It was related that, on the day that was fixed to poll the troops, that French officers, with the aide de camp of General [unidentified], went to the 1st and 2nd Battalions, that they gave money to the soldiers, and that they had received the promise that they would vote for French service, but that the Polish officers were greatly amused and said that one would see how effective this [attempted bribery was] when one asked for the troop's opinion.

The following day, General [unidentified] went to Neufalwasser to poll the 1st and 2nd Battalions. We there opposed much less that we knew, since for a long time, the general was hated as much by the troops as by the officers, and that his presence produced an effect contrary to his views. Prince Radziwill and I went to the 3rd and 4th Battalions, which were on Heubuden Island. First, we had the officers [to vote by signing] the list of names, formed in two columns. All the French signed for service with France and all the Polish signed for service with Poland. The non-commissioned officers and soldiers of the two battalions then voted in the same manner.

We drew together the vote of the four battalions: the 1st and 2nd voted the same as the 3rd and 4th Battalions and all the pieces relative to this operation were sent to Governor Rapp.

These events greatly disturbed the French officers of the Legion. They contained, in their ranks, about 50 French émigrés, who saw, in the transfer of the Legion to Warsaw, a second expatriation. The reformed officers of the old Army of the Rhine saw, further persecutions that they suffered as a result of their association with Moreau and Pichegru in this movement. It was with much difficulty that I was able to assure them that the words emanating from officers assured their recall to French service. I told them:

> Why do you want, from now on, and without having received the order, abandon the brave men with whom you have served and who you have placed in the French line army? If we quit the Legion, it will pass into the hands of the Polish officers in it and we will lose to the King of Saxony the bravest and most numerous of his Polish Army, without this loss being profitable to any other nation. The French officers of the Legion should, therefore, remain with their troops until they have received orders to leave. The calls that they made can only be considered a request to remain in French

service, which, if one wishes, can be seen as a positive rejection of service in the Polish Army. I say to you, what concerns me is that I have decided not to abandon my position until I have received an order to do so.

My representations did little to calm spirits and we awaited the events of the vote.

Around a month later the Legion was ordered to leave the villages surrounding Danzig (2 September 1807) to move to the Polish Army. It had to take everything that belonged to it with it. The French officers made many complaints, but it was necessary to obey orders, or to renounce their positions. The Peace of Tilsit was executed. Nearly all of the complaints were without redress, and it would have been dangerous for the officers, in this case, to submit their resignation after having ended such a brilliant campaign as they had just finished. They would have overloaded the Counselor of State with their combined rights or customs; and, again, one would have suspected these officers had been forced to submit their resignations. Such were the new representations that I made to them and none of them abandoned their posts.

The Legion arrived in Posen, in the Duchy of Warsaw, where they were placed in Dąbrowski's division. Prince Radziwiłł had gone to Warsaw, because he did not wish to find himself under the orders of this general and I was obliged take a lodging, near me, for a Polish officer to act as an interpreter.

When I presented myself at General Dąbrowski's headquarters, he pranced about his headquarters, laughing aloud. I recounted to him the Legion's actions before the enemy; that it had gained a reputation that could only honor the Polish Army, and that I was content to see it witnessed by the gaiety and the pleasure with which he had seen it placed in his divisions. After this, the general became very serious, but welcomed me very well. I would not have, certainly, suffered him to behave otherwise, in my regard. I thought to still see his face covered with the curses of the poor wretches of Dirschau and it appeared to me that his lameness, very visible when he danced about in my presence, was the beginning of this punishment that Moses threatened so many times to those who did not observe the commandments of God. I already knew that Marshal Davout, Governor of the Duchy of Warsaw, had condemned to him the considerable restitutions. I disliked this Polish general, as much as I disliked General [unidentified) and I took, in his regard, a very decided tone, all in accord with that due to his grade. Dombrowski was of a great stature; he was plump, he had a large, oval face, his nose and eyes were large, and his forehead was large and uncovered.

During the whole period that the Legion remained in Dąbrowski's division, the general did not stop to torment it, causing it run from cantonment to cantonment. Général de brigade Kosinski, a man of small stature and small face, perfectly seconded his commander, Dombrowski. The captains and commanders of the Polish regiments urged our soldiers to desert so that they could incorporate them into their regiments so that they could return to their lands, those soldiers who were their serfs, because, in Poland, the peasants belonged to those who owned the land they cultivated. However, the soldiers of the Legion placed themselves above those of the other regiments of Dombrowski's legion. The Legionaires were incomparably better equipped than them and all efforts by the commanders and officers of the Polish regiments were useless, as they feared being returned to the land as serfs, and this, alone, was sufficient to keep them from deserting.

The French officers of the Legion raised great cries and they tormented me that I request their return to France. It was useless, as I exhorted them to wait for orders from the Emperor, to whom several requests had been sent for their recall. However, the complaints of these officers were well-founded. The civilization of the Polish gentlemen allowed them to retain a bit of savagery in their tone, as well as their habits, and a few things that were incompatible with generosity, frankness, playfulness, human processes, and the politeness that one finds among the simple bourgeois of France. Only those Polish officers who had traveled to France knew the brilliant qualities by which the French distinguished themselves from other people.

The events that I am about to recount show how distant the Poles were from the French; and I speak only of the men, because the Polish women, in their civilization and instruction, were perhaps higher than those of the greatest capitals of Europe, and for their physical beauty, the whiteness of their teeth, the brilliance of their eyes, and their figures. One found there many Jewish women with round figures that were very pretty, proving that the races of this nation, originally from Palestine, were mixed.

General Dombrowski, old and lame, was about to marry a young lady from Posen, a city where we still were in the month of October 1807. The Polish ladies loved to see the troops and, when an officer wished to please them, he showed them the men he commanded. It is to this end that General Dombrowski had the units assembled that formed his division and the Legion was included in that number. He ordered that it perform the manual of arms to a general command, which was given to me, in French, by an officer that I had near me. The time that was necessary for me to explain the command in Polish, added to that to give this command, resulted in it being delayed [in relation to the other units] and the manual of arms, in the Legion, was, necessarily, executed after the other units in the division. An *adjudant général* came to tell me to have an officer repeat the commands in Polish. "Go tell your general," I told him, "that my troops do not understand any commands, except those coming from me!" It is pointless that I repeat the other bitter explanations that I had, both with Général de Brigade Kosinski and with the *adjudants généraux*. I easily saw that the colonels and the other officers of General Dąbrowski's troops considered the Legion as less well drilled than their own and I profited from a quiet moment to say, in a group of generals and commanders of horse units, "that I would open a school of practical instruction, when it was desired, and I would serve the officers and the soldiers of the Legion, and from there, as before the enemy, one could see what we were in a state to do; that those of these gentlemen who had served France should know that they had been treated with generosity; that the French officers of the Legion, most of whom were gentlemen[35], had not come to the Polish Army to take positions that belonged to the nobility of the country; that I would have Prince Radziwiłł summoned to come, so that he would arrange the situation, as he wished, with his compatriots. As for the rest, I would explain myself most amply, to General Kosinski after the exercise."

I went to General Kosinski after having returned the troops to their lodgings. He had left the exercise ground at the same time as I and [when I arrived] he had already changed into an ample silk dressing gown, covered with large flowers, and that dragged on the ground. It resembled those that are family heirlooms in some of the old German chateaus. This robe hung in a bizarre manner on his muddy boots, adorned with spurs.

"You have served in France," I told him, "and you know that our generals have

---

[35] Editor: By saying they were "gentlemen," the author is implying they are of better families, not bourgeois or commoners.

never obliged the Polish officers to command their troops and to exercise them in French. One had still less required that these troops know the French maneuvers before they had learned them. You know that I had said, on the field, where the division assembled a half hour ago and it is pointless that I repeat it. I have come here to tell you that good order exists in the Legion or ask that General Dąbrowski give me a written order to turn over command of the Legion to the Polish commander that he designates, or that he issue a new order, joining these troops to other corps in his division, that it might exercise on parade. The French officers of the Legion have led it here, as ordered by the Emperor, but they are not in the service of Poland and one cannot oblige them to serve Poland, and one cannot oblige them to speak the language of your nation. However, all the orders that I receive are in Polish and this is less necessary, for your staff, where the general and officers speak only French. You have paid us one month's salary and you have noted our maneuvers in a manner that is humiliating for the officers who have the honor of leading it. I will submit to you in writing what I have said in person, if you desire it, but I advise you that I am going to make it known, word by word to authorities, so that the legion will be treated with the greatest justice. I owe this to the officers and to the troops, all of whom are honorable men."

General Kosinski appeared shocked and the tone

of voice that I taken towards him, which left him no doubt of the resolution I had taken. He made objections and I repeated to him, several times, that I would not take back a single word I had said to him; that my propositions were presented as I had planned, and that the peace of the legion demanded that I not deviate from this plan.

From this moment, I was given no more orders to muster, and I impatiently awaited the response to the report I had sent to Davout, as well as Prince Poniatowski. It arrived on the day that the Polish soldiers [from Dąbrowski's division], guided by their officers, no doubt, intoxicated some of the Legion's soldiers. This caused them to take off their regimental coats [i.e., abandon the unit]. However, the following day, our soldiers returned and related what they had done. I was not prevented from taking the action that I thought proper to prevent this new recruitment [from the Legion by other Polish units].

The master at arms of the Legion, at the instigation of a few officers, moved to the encampments, where he distributed blows with his hands and kicks to all the soldiers of Dąbrowski's division that drank with ours, who then drew their sabers to obtain satisfaction. However, this action was not known by Dąbrowski's division. Blows with sticks quickly followed the punches and the bulk of our troops entered the brawl, with Dąbrowski's troops falling before ours, and they hid themselves. One no longer saw soldiers from the Legion in the streets and it was necessary to send the corps into the villages a short distance from the city, where one could not do wrong by them. Degasq, a French gentleman, the drum major of the Legion and master of arms, was charged with directing the expedition that I have just related. General Dąbrowski had called a council of war, which had condemned him to death. Pallandre, captain of grenadiers, disguised him, so he could take the road to Frankfurt-am-Oder, to rejoin his old regiment.

It is probable that he had never distributed so many blows to a division in a single day before. It is also true that never had a single regiment decided to beat an entire division with sticks. Prince Radziwiłł related to me, subsequently, that this brought much amusement to Warsaw, and that, for a long time, in all of Poland, everyone spoke about the beating of Dąbrowski's division.

This was because this general dreaded the jokes that would inevitably arise as a result of his calling the council of war. Degasq was nothing but a superb young man, a French conscript, who had received a good education, was very polite, spiritual, and conducted himself well. His only fault was being master at arms, like 152 other non–commissioned officers and soldiers that he had led in the beating escapade, which further increased the pride of the Polish soldiers of the Legion.

This affair was not the first correction of this type that the Legion gave out. I speak of that of Thorn, through which the Legion passed en route from Danzig to Posen. An hour after our arrival in Thorn, where was stationed the French division of General Gudin, who I knew well, one came to me to tell me that 100 soldiers from the Legion and Gudin's division, with sabers in hand, had fought it out, with many dead and wounded. I ran to General Gudin's headquarters and asked him to sound assembly, before we had been informed of the reason for the quarrel. The reason for the quarrel was as follows: "There were 153 French masters at arms in the Legion," I told him, "and you can see that your division does not have an equal number to oppose them!" I called Captain Pallandre (who I considered the chief of the masters at arms, in order to know of their conduct, in all circumstances) and he informed us that, at the moment when the Legion entered the city, they encountered French grenadiers on the bridge, who made unbecoming comments, not believing that the soldiers of the Legion would understand them.

However, our masters at arms had noted them and, after our troop entered their lodgings, they went out to demand satisfaction. The assembly was then sounded. I sent Captain Pallandre to the masters at arms to obtain from them the promise that the affair was ended, and General Gudin went, personally, to his assembled troops. Gudin had difficulty believing that Pallandre's action was sufficient to assure the Legion's behavior. I explained to him why all other actions were pointless, and I went to return our companies to their lodgings.

The Legion was still in Posen when I received, from Prince Poniatowski, Minister of War of the Grand Duchy of Warsaw, the order to lead it to Meseritz, a small city situated between Posen and Frankfurt-am-Oder. Prince Radziwiłł joined us while we were in route. We remained there for a month. We were alone there and General Dąbrowski did not send us any orders.

The Legion was then sent to Gniezno, eight leagues from Posen. Prince Radziwiłł returned to Warsaw.

The Legion was still at Gniezno, when the order came to have the troops swear allegiance to the King of Saxony. This was most embarrassing for the French officers of the Legion, whom I had already asked to be returned to France. These officers asked me what they should do, but I could give them no advice. I advised them simply, that as far as I was concerned, I would not refuse to swear the oath to the King of Saxony for the time being. It was pleasing to the Emperor to leave me in the Army of Poland and that this army would continue to be under Maréchal Davout, Duke of Auerstädt, until my recall to France. Heads heated at the oath-taking. The soldiers, discontent with the aggravations that they had suffered under General Dąbrowski, lined up alongside of the French officers, and said that they would not take the oath. They accused the Polish officers of having tricked them with false promises when they were asked to vote for French or Polish service.

Prince Radziwiłł, by his conduct, didn't dare to appear before the Legion, because it was on his part that the Polish officers had made the promises that had determined them to vote for Polish service. For a long time, I called on him to return and appear before his compatriots, because, in Poland, the families with great names and fortunes have great influence over the ordinary nobility, because of the influence they exercise on the government by their clientele and those they employ. I understood, from General Kosinski that Prince Radziwill greatly feared seeing me spontaneously quit the Legion, with all the French officers under me. This would greatly embarrass the prince since he was incapable of commanding the Legion without me.

The day of the oath arrived, and the 1st and 3rd Battalions were with me.

I had the 1st and 3rd Battalions assembled on a field near Gniezno. General Dombrowski had sent several Poles to observe the administration of the oath. He had, doubtlessly, wished to see me make errors, but I took measures to not give him this satisfaction. As I knew that the officers of the 1st Battalion had refused to take the oath, I believed that I should first ask the non–commissioned officers and soldiers of this battalion to take the oath. I had the battalion formed in square, with the soldiers' facing inwards. The commander of Gniezno read the oath, in Polish, and then invited the troops to submit to it by raising the hand and extending a finger into the air. But this signal, which he did himself, was not repeated by a single person. I asked the commander to speak to the troops and he harangued them without success. Then, I thought I should invite the Polish lords, who were present, to pass before the battalion and to say to the troops that the oath asked them was only the result of the vote they had given at Danzig, to enter into Polish service. Finally, the soldiers raised their hand and extended the index finger, but they did it despite them. I asked, then, that the oath be taken by the non-commissioned officers, but they acted like the officers, and each did as he wished. All the 4th Battalion took the oath. The French officers and non-commissioned officers of the 2nd Battalion refused the oath.

I should remark, on the subject of this oath, that, at Danzig, the Poles of the Legion were eager to vote to serve their country and that their refusal at Gniezno, under the circumstances of which I have spoken, was in contradiction to the first vote. If I had not taken some steps to remind them that they should be in accord amongst themselves, the generals and the commanders of Dombrowski's division would not fail to blame me of having influenced the troops and to have provoked the insurrection and the disorganization that would necessarily follow the rejection of the oath. The refusal of most of the French officers would have no consequence, but, as I found myself the only superior officer of the Legion who was present, not a single soldier would have taken the oath if I had refused to do so and this would have made me responsible for everything that followed the refusal. I recalled, in addition, that this was the response made to Prince Radziwill by General Rapp, governor of Danzig, who had sent the Legion out of the French Army. My secret thoughts had been that if the Emperor had desired that this corps remain in his service, it would have been pointless to ask his opinion on this, as the heart of the Legion was French.

While all these events occurred, Prince Michel Radziwill held himself apart from the Legion. He had taken care to leave Warsaw and move to Posen during the oath taking ceremony, but he soon appeared at Gniezno, when it was completed. He then left to return to Warsaw. I believe that his conduct, in this situation, was in accord with the politics of his family.

I believe that Maréchal Davout was tired of the complaints the Legion directed to him. He was informed of everything that occurred with it. I presume that he agreed with Prince Poniatowski, the Polish Minister of War, that we should be separated from Dombrowski's division, because we received an order to move to Wresnia and the surrounding villages. Wresnia is a small city on the road from Gniezno to Warsaw. It belonged to the Poninskis and was almost entirely populated with Jews. I then found myself sheltered from the harassment of Dąbrowski and his agents. As a result, good discipline was re-established in the Legion. It conducted itself well towards the local inhabitants and it conducted itself better than the Polish regiments, which the nobles recognized, themselves.

We had detached, to Warsaw, Captain Count de Chavannes, a former chamberlain of the last king of Poland and Knight of Malta Machinfort, also a captain in the Legion. It was through these men that I learned what was said about us in Warsaw society.

I did not want to leave the Legion before it was better trained in exercises and maneuvers than the Polish regiments. I had examined Dąbrowski's division, and I saw my superiority over the corps commanders in it. I gave detailed orders to follow the particular methods that I had developed, and I took measures to put them into practice. The 3rd Battalion was with me at Wresnia.

The three other battalions were informed of the progress of the battalion under my supervision, and, with the exception of the 1st Battalion, which was too far from me, they trained so as to not fall behind the 3rd Battalion. Prince Radziwiłł also came to Wresnia and he assisted in my lessons. I engaged him to serve as an instructor and I proposed to give him lessons in my room, but he was more attached to his guitar and his pipe than in the regulation of maneuvers and exercises. As a result, he learned nothing at all. It was his opinion that, if the princes and great lords were obliged to work like the men of a lower station in life, then birth was a oppressive burden instead of an advantage, independently of the obligations and the genius which was attached to greatness, if it was necessary that he study, tormenting the spirit and body as one needed to do to learn.

At the beginning of 1808 the Legion received orders to move to Zakroczym, 10 leagues from Warsaw, on the road to Plock. There, it occupied cantonments along the Vistula, over a distance of 40 square leagues. The Legion continued its good discipline.

The Emperor recalled the French officers of the Legion to France. I was required, in preparation for their departure, to organize two detachments, so as to not to encumber the route. I first had the lieutenants and sous-lieutenants depart and I held the company commanders until they issued reports on the unit's linen and socks, in accordance with the Regulation of 8 Floréal, Year 8.

It was in Warsaw that I was to choose sixteen officers and non-commissioned officers, who were to receive decorations.

As I did not speak Polish, I could not properly prepare the recommendations. I was obliged to summon Prince Radziwiłł to come. We succeeded in completing the work. The recommendations were forwarded to the Minister of War, and it was as a result of this that we received the Polish decorations of which I have already spoken. Prince Radziwiłł and I were not included in the list, but we each received a 1st Class decoration.

Some days later, I received orders to reduce the Legion to two battalions and to use the excess to form a fifth Polish infantry regiment. The departure of the French officers left only the officer cadres for two battalions. Only one of the French officers, Captain Lemaire, remained in Polish service, since he had many debts in France, and three months after the departure of his comrades, he died of shame. All the French officers of the Legion, after their return to France, were placed in line or light infantry regiments, except for the Chefs de bataillon Vanrose and Roumette, and Quartermaster Treasurer Lebrun. However, they were employed in Spain. I remained in Poland, as authorized by Maréchal Davout, the French officers who were members of the administrative council and the accounting officers. It was necessary to send a report of the administration of the money and material of the Legion, both to the French and Polish governments. The work was difficult, as the changes in the Legion had been immense and that it had had no peace until it occupied its cantonments at Zakroczym, near Warsaw.

Finally, during the course of March 1808 the Legion became the 5th Infantry Regiment of the Polish Army and received orders to move to Warsaw. Prince Radziwiłł came to Zakroczym to enter into the capital of his country at the head of his corps. It mattered little to me that I retained command as I had before, or to momentarily give it to Prince Radziwiłł as he entered Warsaw. My reputation had been made in the Polish Army and throughout the Duchy. It was known, everywhere, that the Prince had still not learned his trade and I had no need to make this evident in the capital. On another side, I did not wish to deprive the Radziwiłł family, which was assembled in Warsaw, of the pleasure of seeing this prince as the commander of the only Polish corps, which had achieved such a wonderful reputation in the war that was just ending.

During the march to Warsaw, he constantly asked me to stop, so as to enter the city at night, because of the large number of people who had gathered on the streets through which we must pass, to see the Legion. I thought that he feared that Prince Poniatowski would order us to maneuver, as, in this case, he would only be a spectator. I told him that we were going to make a triumphant entry, before the eyes of his family and friends, and that he would present himself as the commander of the regiment. He informed me of the size and shape of the Saxon Square, on which we would be formed into line. I then told him the commands he was to issue and that I would stand by his side to prompt him in a low voice. Night began to fall as we arrived at Saxon Square and the regiment formed itself in line, as I shall relate. There were many people there, as had been in the streets through which we passed to reach the place. Prince Poniatowski, with all the Polish generals, had awaited us for a long time.

Prince Poniatowski was unable to review the regiment until the next day. Many people were to be found on the square. The Polish Inspector General of Reviews, with a general, read a proclamation to each company, to remind the troops that they were in Polish service. I gave my arm to Princess Radziwiłł, mother to Prince Michel. Together we made a review of each company. There was much snow and ice on the square and the temperature was cold. This princess had much spirit and was well educated. She spoke seven or eight languages. For a long time, she had been the first lady of the Russian court. It could be said that it was she who had given this court, under the Emperor Paul I and Empress Catherine II, the tone that still reigned there.

She appeared pleased as, when we approached the company at the head of the column, when I ordered their captain to have them present arms. The Princess was such a physiognomist that she judged at a glance; from which nation each soldier came. She discovered, also, on the face of each man, the country in which they were born. She spoke to each in the language of his birthplace and the responses that the soldiers made to her proved that she was correct. She judged that Lieutenant Paly-Rache was a Jew and that he had made extraordinary pretensions, without being able to determine what they were, and she was correct. He said he was a prince of the House of David. He carried on his shirt, and under his vest, a unique plaque to which the Polish Jews had great veneration.

I had seen, in Danzig and in Zakroczym, many hundreds of Jews render him the honors that they said was due someone of his rank and it sufficed to show them his plaque to have then prostate themselves before him. He spoke and wrote eleven different languages. He was very highly educated. He was able to preach in all eleven languages. He had been a doctor in the armies of France, professor of several sciences, etc. He was extraordinarily brave. He had been wounded many times while serving in the Legion and he was a stunning warrior.

He was a very quick teacher, and, in a very short time, he taught Arabic to Madame Muy[36], the wife of the French general of this name. He was the most sober man I have ever known. He had taken service in the Legion so as to go, as a soldier, with the French Army, which, it was said, was about to make an expedition to India by marching past the eastern side of the Black Sea. His project was to raise an army of Jews to allow him to recapture the throne of his ancestors.

An adventurer, who I had taken into the Legion as an officer in Magdeburg, at the time where I had almost no officers, and who presented himself as the Baron de Mayern, followed the Legion until it crossed the Oder. He said he had been an Austrian officer. He remained in the rear at Stargard, eight leagues from Stettin. I learned, subsequently, through the stragglers who rejoined the Legion, at the beginning of the siege of Danzig, that he had gathered up 150 to 200 men of the Legion and that he had proclaimed himself king of a small, wooded canton near the Vistula. His capital was a village and he had organized a military force in his states. He guarded the soldiers of the Legion who wished to remain with him, and he treated generously those who did not wish to tie their fortune to his. As French detachments passed through his kingdom,

---

[36] Jean-Baptiste-Louis Philippe de Félix, Count du Muy (1751-1820), lieutenant general in 1792, he was charged, in 1806, with the government of Silesia. It was then, perhaps, that he met the Jewish officer, Paly-Rache.

those men left his military organization and rejoined the Legion during the siege of Danzig.

I met with him and spoke to him of his kingdom. He made his responses laughingly and assured me that he had effectively united a number of soldiers to allow him to avoid being killed or captured by the von Schill Freikorps. I caused him to observe that it would be inconvenient for a sovereign to be a simple lieutenant in the Legion and that I could only receive him with consideration of how well he had treated our soldiers who had rejoined us and, furthermore, to thank him that his soldiers had not joined von Schill. I left him free to return to his states, but I counseled him to not speak of his royalty. I never saw him again.

Baron Mayern had much spirit, he spoke many languages, and he explained himself, with ease, in French. He was around 30 years old, of short stature, lively and very agile. It is assured that von Schill had made propositions to him. The 200 men of the Legion that he had gathered up were a good beginning to make a partisan war, as did von Schill. He could have made his military fortune by placing himself in the corps commanded by this audacious warrior [von Schill], but, perhaps, the soldiers of the Legion wished no part of it.

I liked him enough to keep Baron Mayern in the Legion, because men of his character are carried to fortune by bursts of action, but some officers who knew him during the march from Magdeburg to Stargard, did not speak well of him. I thought, after this, that I should not receive him.

Many other extraordinary events occurred in the Legion, of which I shall also speak.

There were several suicides and a number of other adventures of this nature. For example, Kuhn, the surgeon-major of the Legion, hung himself, while at Thorn. The Marquis de Cravey, a lieutenant and returned émigré, drowned himself after being rejected by a woman. The soldiers drowned one of their comrades who was ill with punais[37], under the pretext that his sickness was incurable and that those who are attacked with it are a charge to himself and to others, and that, in Poland, it is a generally established practice among Polish peasants to drown them.

A Polish soldier will not give assistance to another soldier who has an accident, for example if a soldier falls into a ditch filled with water, and from which he cannot escape without the assistance of another. This proves, without a doubt, that the Polish serfs are too stupid to understand the principals of the Gospel and from that it is impossible to credit them with the Christian sentiments of charity, hospitality, and goodwill, which one finds among the people of France.

These observations are even more applicable to the Russian peasants, since they are more enslaved and more stupid than those of Poland. Platon, himself, thought that he must have slaves from the warm country he inhabited and that the property owners of Poland, as well as those of Russia, thought that the cold, in their countries, to which those who cultivated the ground must become accustomed, by force, to withstand the rigors of the winter. At least, this is what I was told by many educated Poles.

In the French Army, it is forbidden for soldiers to remove the wounded from the battlefield and lead them to the flying ambulance, because the concerns of this type diminish considerably the battalions. However, among the Poles, such a prohibition is pointless.

The causes of this lack of charity are, on one part, because it is in the interest of the masters to conserve the

---

[37]Editor: "Punais is an obsolete term, meaning "stinking." The probable illness is chronic atrophic rhinitis, which is a fetid inflammation of the nasal mucosa, a family of diseases that emitted a terrible odor and made social life impossible. Victims frequently committed suicide. Drowning someone with this disease is not out of the question.

life of their serfs and, after a battle, all who might recover are removed from the battlefield. This is done by the serfs, but only on the express command of their masters.

The particulars that I have recounted, in explaining the basic organization of the 5th Regiment of the Polish Army, prove that, when the Legion was formed, all the adventurers sought to join it and it is in part, for this reason that the corps of this type does not last long. An officer is not good, except when he has a reputation to defend. It is the same thing for the non-commissioned officers and, I say, that the reputation of an officer or non-commissioned officer is applicable to an entire corps, since the reputation of this corps is formed of the reputations of the individuals who form it.

This proposition is true, since our old regiments remain good, no matter how great the losses that they suffer before the enemy. It suffices that there remains a certain number of veteran officers and non-commissioned officers to give the new soldiers that join the regiment these same sentiments of honor and spirit on which rests the regiment's reputation.

There were, in the Legion, around 50 returned émigré officers and I was happy with them and they, certainly, were happy with me. There were other French officers who had marked our political troubles. There were officers on half pay[38] who returned to service, Indians, Americans, Russians, Jews, Poles, Vendéans, Germans, Swiss, Italians, and others. Captain de Villeneuve had <u>been a general</u> of the Chouans[39] and, in this capacity, he

---

[38] Editor: In the list of officers at the end of this manuscript, these officers have been described as "returning from retirement."

[39] Editor: The Chouans were rebels in the Vendée who rose up against the Revolutionary Government beginning in 1794 and fought a bloody and ruthless civil war. The name "Chouan" was, for a long time, an insult in France. It was synonymous with brigand, and it is true to say that the conduct of several men bearing this name justified this synonym, but they were guerillas. The Chouan Royalists confess, to this day, that there were smugglers, highwaymen, pillagers, and professional assassins found in their ranks. Four brothers, named "Cottereau," are considered to be the founders of the Chouannerie. Smugglers before the Revolution, they lived in the village of Saint-

had fought against the French government. The Count du Puget, a very brave man and good chemist, had been a colonel in the French Army before the Revolution, and now was only a captain. The Count de Chavannes, the last offspring of the illustrious family of that name, had served as a colonel and chamberlain to the last King of Poland, and was now only a captain, even though he wore the ribbon of the Order of Saint-Stanislaus. Spruglin, of Swiss origin, and a man of much spirit, had been the Swiss ambassador to France. He served as a captain and was an excellent soldier. There were veteran Knights of Saint-Louis and of Malta serving in the same capacity. The opinion of the émigrés was that they had only taken up arms against the Revolution to support the monarchy, and that this form of government had been re-established in France[40] so they volunteered to join their compatriots for the prosperity of France. In truth, our political troubles had occasioned a change in the dynasty, but such a change was not extraordinary. Royal lines were extinguished, like men, in every known land. This was in accordance with the laws of nature. The party, alone, was stable and that it was only to it that one attached oneself, as much it was the only nourishing mother of all its children. Besides

---

Ouen-des-Toits, near Laval. It is said, in order to recognize each other in silence in the dark woods, and to avoid being surprised, they used the lugubrious cry of a nocturnal bird known as a "chouette" or "chathuant." It also appears that, before the Revolution, they were known as the Chouans, which is the Avigne or Breton pronunciation of the word "Chat-chuant." Whatever might be the etymology of this word, the brothers Cottereau saw themselves ruined at the time of the Revolution by the abolition of all old fiscal laws that made smuggling profitable. They became the enemies of the new regime and embraced the cause of a government which would have hung them during its existence, had they fallen into its hands. They formed, in 1793, near Laval and Gravelle, an assembly, initially formed of other smugglers, to whom they were soon joined by malcontents of the neighboring countryside. Jean Cottereau, called the "Chouan", the oldest of four brothers, was chosen by the bands to be their commander and the entire group took the generic name of the "Chouans", which subsequently was given to all the Royalists in Brittany and part of Normandy. They were wiped out at Quiberon on 20 July 1794.

[40]Editor: The re-established monarchy is a reference to Napoleon's becoming Emperor.

the pretender, known as Louis XVIII by some, and as the Count de Lille by others, had allowed them to go back and to serve France. This had cleared them of the oath of loyalty that they had sworn by asking them to hold their hearts for him, and, in this, their honor was maintained.

The Legion only remained in Warsaw for three days and then left to move to Kalisch as part of the division commanded by General Zajączek, and to there take its position as the 5th Polish Infantry Regiment. I had to remain in Warsaw with the administrative council and the officers charged with the details. It was entirely French, since General Zajączek, in beginning the organization of the first battalion at Landau, before the arrival of the Polish officers, could only form a French council. The opinion of this general was, otherwise, that his compatriots were little able to direct a regular administration, and, above all, that they should be closely supervised, and their work verified.

During my sojourn in Warsaw, I continued to correspond with General Zajączek. I sent him my written opinions on the Polish Army and the means that could be used to perfect its organization to achieve the necessary instruction and to preserve its administration from the torturous regulations that so distressed the French soldiers. He informed me that he was stunned that I so well knew his nation.

General Zajączek had desired that I remain in Polish service. He had the project, according to what he had confidentially told me, to have me named general major of his division. He mentioned that Major of Artillery Pelletier, who was entering Polish service, did not have, as I did, the rights acquired by the recognition of the Polish nation. I was told that I had the rank of colonel, according to the practices observed by the military powers of the continent, with regard to soldiers who passed from one service to another, and that my brevet had already been sent. But these wonderful hopes were not enough to cause me to leave French service, where I had acquired the rights of 28 years of service.

Instead of accepting the rank of a Polish general major, a grade which was given to Major of Artillery Pelletier, shortly after he accepted service in the Polish army, I asked for employment as Under Inspector of Reviews in the French Army. I had suffered frozen feet and legs during the winter of 1806-7. I marched with much difficulty, and I had other infirmities that prevented me from active service in the infantry, the arm in which I had always served.

I sent my request from Warsaw on 13 May 1808 and on the following 6 June I was named Under Inspector of Reviews by a decree issued in Bayonne. These two cities were more than 600 leagues from each other, and the postal system had delivered it with astonishing quickness that the response should have come from Paris in such a short period. It had remained there for several days, in order that the Minister of War had time to join to it a report and to forward it to Bayonne, where the Emperor was at that time.

List of officers of the Légion du Nord, by rank and by order of their arrival in the corps, with notes:

Zajączek, Général de division, 1 October 1806, ancient colleague of Kosciusko, in the national war to prevent the partitioning of Poland.

Vanrosen, Chef de bataillon, 18 November 1806, a Belgian with service in India.

Coqueugniot, Major, 1 December 1806, noted in the memoir.

Roumette, Chef de bataillon, 23 Decembrer 1806, noted in the memoir.

Stokowski, Chef de bataillon, 4 March 1807, Pole with Russian service.

Martres, Chef de bataillon, 31 March 1807, French coming from retirement (half pay)

Jeromski, Chef de bataillon, 1 April 1807, Pole coming from his lands.

Prince Radziwiłł, Colonel, 23 April 1807, Pole as noted in memoir.

Count Sobolewski, Colonel, 28 April 1807, Pole as noted in the memoir.

Golaszewski, Chef de bataillon, 19 May 1807, Pole with French service.

LeBrun, Quartermaster Treasurer, 11 November 1806, retired French with service in India.

Trautner, Adjudant-major, 15 October 1806, French coming from retirement.

Mougenot, Adjudant-major, 26 October 1806, French from the Isambourg Regiment.

Hoppen, Adjudant-major, 23 October 1806, Lithuanian with French service.

De Vaisvres, Adjudant-major, 1 November 1806, French, returned French émigré with service in India.

Hermann, Adjudant-major, 20 August 1807, French coming from retirement.

Grabinsky, Captain, 10 October 18065, Pole with French service.

Vernier, Captain, 18 October 1806, French coming out of retirement.

Poupart, Captain, 28 October 1806, French coming out of retirement, lawyer.

Larrey, Captian, 30 October 1806, French coming out of retirement from Paris.

Pleux, Captian, 1 November 1806, French coming out of retirement from Paris.

Count de Chavanne, Captain, former colonel, 8 November 1806, French, former chamberlain, decorated with a crachat.[41]

Duboishaumon, Captain, 15 November 1807, French, returned émigré, former infantry officer.

Caillet, Captian, 15 November 1806, French coming out of retirement.

Brondelle, Captian, 18 November 1806, French coming out of retirement.

Henry, Captain, 19 November 1806, French coming

[41] Editor: Crachat does not translate. It is a "decoration of a superior grade" according to Petit Robert. It must, therefore, be a rosette or some such device to indicate a higher grade of some order.

out of retirement.

Mordret, Captian, 21 November 1806, French coming out of retirement from Paris.

Scalabrino, Captian, 29 November 1806, French coming from the National Guard.

D'Andlau, Captain, 29 November 1806, French coming from retirement and the National Guard.

Lemaire, Captain, 29 November 1806, French coming from retirement and the National Guard.

Lesire, Captian, 29 November 1806, French coming from the National Guard.

Pallandre, Captian, 29 November 1806, French, director of a military hospital.

Thouars, Captain, 30 November 1806, French without state, former Jacobin, I believe.

Grosourdy de Saint-Pierre, Captian, Knight of Malta, 24 December 1806, French, returned émigré, veteran officer.

Majewski, Captain, 30 December 1806, Pole having French service.

Warnery, Captian, 1 January 1807, Swiss and officer in Swiss service.

De Beaumont, Captain, 1 January 1806, French, returned émigré, no previous service.

Sprunglin, Captain, 1 January 1806, Swiss and former ambassador to France.

Godlewski, Captain, 10 January 1807, Pole with French service.

De Mallard, Captain, 10 January 1807, French, returned émigré, veteran officer, old.

De Junemann, Captain, 31 January 1807, French, returned émigré, veteran officer.

De Saint-Martin, Captaine, 1 February 1807, French, joined only to be decorated.

Pouciatka, Captain, 10 March 1807, Pole coming from his lands.

Larocheskalki, Captain, 10 March 1807, Pole and engineer, he had little service.

Veggi, Captain, 13 March 1807, Neapolitan and Knight of Saint-Janvier.

Macschehy, Captain, 13 March 1807, Irish coming from the legion of the same name.

Thérondel, captain, 19 March 1807, French, returned émigré, no prior service.

Bellanger, Captain, 19 March 1807, French coming out of retirement.

Count de Puget, Captain, former colonel, 19 March 1807, French, returned émigré, intrepid as he was old.

Lelidec, Captain, 19 March 1807, French, returned émigré, former officer.

Grand, Captain, 19 March 1807, French coming out of retirement, educated.

Hobacq, Captain, 19 March 1907, French, returned émigré, veteran officer.

Mayern, Captain, 25 March 1907, German or Swiss, coming from foreign service.

Bitchfeld, Captain, 30 March 1807, French, returned émigré, coming from foreign service.

De Tardivel, Captain, 31 March 1907, Indian with service on land and sea.

Perrin, Captain, 19 May 1807, French coming out of retirement from Paris.

Mislotawski, Captain, 10 March 1807, Pole coming from his lands.

De Bohn, Captain, 16 Mach 1807, Dutch, coming from Austrian service.

Machinfort, Captain, Knight of Malta, 1 June 1807, Italian, former political agent.

De Monjavoust, Captain, 18 July 1807, French, returned émigré, former officer.

De Zévallos, Captain, 1 June 1807, French, returned émigré, former officer.

Besnard, Captain, 21 July 1807, French, returned émigré, former officer.

De Villeneuve, Captain, Commander of the Order of St.-Louis, 1 August 1807, former Chouan general.

De Rècalde, Captain, 16 August 1807, French, former aide-de-camp to the Duke d'Enghien.[42]

Marquis de Cravey, Lieutenant, 20 November 1806, French, returned émigré, former officer of the army of the princes.[43]

Belhomme, Lieutenant, 29 November 1806, French, entered the Legion as a surgeon.

Le Corrège, Lieutenant, 24 November 1806, French, returned from retirement.

Jacob, Lieutenant, 24 November 1806, French, Alsatian and volunteer.

Doumenc, Lieutenant, 24 November 1806, French volunteer.

D'Aucourt, Lieutenant, 29 November 1806, French volunteer.

Ordran, Lieutenant, 29 November 1806, French, came from the National Guard.

Hartmann, Lieutenant, 29 November 1806, French,

---

[42] Editor: The Duke d'Enghien was the leader of the Royalist cause. He was kidnapped on Napoleon's orders, tried and executed. It is very surprising that this man was able to return to France and serve Napoleon.

[43] Editor: The Army of the Princes was formed of Royalists and commanded by members of the Royal family, as they attempted to restore the monarchy in 1792-5.

came from the National Guard, former engineer at Mainz.

Le Brun, 29 November 1806, French volunteer.

Hann, Lieutenant, 29 November 1809, Pole, came from the French 1st Artillery Regiment.

De Frèmery, Lieutenant, 29 November 1806, French, returned émigré, no prior service.

De Ficher, Lieutenant, 29 November 1806, French, returned émigré, no prior service.

Stehlin, Lieutenant, 29 November 1806, French, came from the National Guard.

Roger, Lieutenant, 31 December 1806, French, came out of retirement.

Kaminski, Lieutenant, 1 January 1807, Pole recruited in France.

Léonard Cyr, Lieutenant, 1 January 1807, Polish volunteer, student.

De Vabre, Lieutenant, 1 January 1807, French, returned émigré, officer of the Army of the Princes.

Markiewitz, Lieutenant, 1 January 1807, Polish volunteer, student.

De Thibus, Lieutenant, 8 January 1807, Brabançon, returned émigré, young man.

De Bohy, Lieutenant, 1 January 1807, French, returned

émigré, officer of the Army of the Princes.

Tkikotski, Lieutenant, 9 January 1807, Polish volunteer, student.

De Clausolles, Lieutenant, 30 January 1807, French, returned émigré, former cavalry officer.

Valentino, Lieutenant, 8 January 1807, French, administrative employee.

Sitzinski, 10 March 1807, Pole, left his father's home.

Tatzicki, Lieutenant, 26 March 1807, Pole, left his father's home.

Sitzinski, 10 March 1807, Pole, left his father's home.[44]

Kahl, Lieutenant, 10 March 1807, Pole, came from his lands.

Szuccodowski, Lieutenant, 16 March 1807, Pole, with French service.

Avisse, Lieutenant, 17 March 1807, French volunteer, young man.

De Goyer, Lieutenant, 19 March 1807, French, returned émigré, officer in the Army of the Princes.

Hegnault, Lieutenant, 19 March 1807, French, returned émigré, I do not know if he had prior service.

Laskowski, Lieutenant, 25 March 1807, Pole, coming from his father's lands.

---

[44] Editor: This may be a duplicate entry.

Ramauge, Lieutenant, 30 March 1807, French, veteran cavalry officer.

Paly-Rach, Lieutenant, 18 April 1807, Israeli, noted in the memoir as being extraordinary for his talents and his origin.

Orlewski, Lieutenant, 22 April 1807, Pole with French service.

Szinkiewitz, Lieutenanta, 30 March 1807, Russian, coming from Lithuania.

Thiébault, Lieutenant, 31 March 1807, French, volunteer, student.

Noviski, Lieutenant, 10 march 1807, Polish, young man who died as a hero, noted in memoir. He is the Polish Assas.

De Labroue, Lieutenant, 23 May 1807, French, returned émigré, still very young.

Grabinski, Lieutenant, 6 June 1807, Pole, coming from the lands of his father.

De Bagars, Lieutenant, 12 June 1807, French, returned émigré, veteran officer.

De Cargs, Lieutenant, 12 June 1807, French, returned émigré, veteran officer.

De la Mothe Salignac-Fénélon, Lieutenant, returned émigré, veteran officer.

Baron de Montaud, Lieutenant, 20 July 1807, French, returned émigré, veteran officer.

Hubert, Lieutenant, 30 July 1807, French, coming from retirement.

Duranowski, Lieutenant, 20 July 1807, Polish itinerant musician, excellent violinist.

De Mollot, Lieutenant, 30 September 1807, French, returned émigré, veteran Gendarme du roi.

Dreizbinski, Lieutenant, 18 October 1807, Pole, coming from his father's lands.

Alphonse, Lieutenant, 19 October 1807, French, great architect, engineer.

Ruffat, Sous-lieutenant, 1 November 1806, French, student.

Guerrin, Sous-lieutenant, 1 November 1806, French, former infantry sergeant.

Glimbowski, Sous-lieutenant, 24 November 1806, Polish with French service.

Berger, Sous-lieutenant, 1 December 1806, French, employed in administration.

De Pradine, Sous-lieutenant, 1 December 1806, French, came from the Prussian Royal Guard.

Narbonne-Pelet, Sous-lieutenant, [date not listed],

French, son of a returned émigré.

Count Angelini, Sous-lieutenant, 3 December 1806, Italian, young man and traveler.

Van Rosen, Sous-lieutenant, 1 January 1807, Belgian, nephew of the Chef de bataillon.

Kinderszinski, Sous-lieutenant, 15 December 1806, Polish, coming from his father's lands.

Flick, Sous-lieutenant, 15 December 1806, Pole, coming from university.

Schmidt, Sous-lieutenant, 21 December 1806, German, coming from university.

Kupse, Sous-lieutenant, 15 December 1806, Lithuanian, student in Germany.

Sikorski, Sous-lieutenant, 15 December 1806, Galician, former Austrian cadet.

Dranuis, Sous-lieutenant, 9 January 1807, French, formerly a non-commissioned officer in the regiment.

Polowski, Sous-lieutenant, 15 January 1807, Pole, coming from his lands.

Makowski, Sous-lieutenant, 19 January 1807, Pole, coming from his father's lands.

Descazeaux, Sous-lieutenant,, 20 January 1807, French, former lawyer's clerk.

Kosciusko, Sous-lieutenant, 28 January 1807, Pole,

coming from his lands.

Kalinski, Sous-lieutenant, 28 January 1807, Pole, served in French Army.

Bruno, Sous-lieutenant, 1 February 1807, French, former student.

Casse, Sous-lieutenant, 6 March 1807, Swede, former non-commissioned officer in the Legion.

Vicaire, Sous-lieutenant, 6 March 1807, French, decorated non-commissioned officer and had retired, volunteered for the Legion to participate in the war.

Pagnol, Sous-lieutenant, 6 March 1807, French, volunteer and former sergeant.

Gilbert, Sous-lieutenant, 6 March 1807, French, volunteer and former sergeant.

Korwalewski, Sous-lieutenant, 6 March 1807, Pole, coming from his lands.

Kulewski, Sous-lieutenant, 6 March 1807, Polish noble, coming from the Prussian Army where he was a soldier.

Dutroy, Sous-lieutenant, 6 March 1807, French, sergeant major of the 24th Légère Regiment.

Naas, Sous-lieutenant, 6 March 1807, French, volunteer, native of Landau.

Kwialowski, Sous-lieutenant, 6 March 1807, Pole, coming from his family's lands.

Neuhauss, Sous-lieutenant, 6 March 1807, Swiss, volunteer, a committed traveler.

Mayern, Sous-lieutenant, 6 March 1807, Livonian, volunteer, student.

Iocham, Sous-lieutenant, 6 March 1807, Livonian, Professor at University of Erlangen.

Vongrowski, Sous-lieutenant, 10 March 1807, Pole, came from his father's lands.

Grabisse, Sous-lieutenant, 9 April 1807, French, volunteer, student.

Rackzinski, Sous-lieutenant, 1 April 1807, Pole, came from his father's lands.

Evelard, Sous-lieutenant, 14 March 1807, French, student, volunteer.

Danin, Sous-lieutenant, 1 April 1807, Pole, came from his father's lands.

D'Agout, Sous-lieutenant, 6 April 1807, French, son of a returned émigré.

Kreb, Sous-lieutenant, 29 April 1807, French, volunteer from Landau.

Legazinski, Sous-lieutenant, 25 March 1807, Pole, came from his father's lands.

Kaiser, Sous-lieutenant, 6 June 1807, Swiss volunteer, no prior service.

Vroblewski, Sous-lieutenant, 1 June 1807, Pole coming from his father's lands.

Laskowski, Sous-lieutenant, 10 March 1807, Pole coming from his father's lands.

Zaiguelius, Sous-lieutenant, 8 January 1808, French, came from the Isambourg Regiment.

Möller, Sous-lieutenant, 5 June 1807, Swiss volunteer.

Guibourg, Sous-lieutenant, 19 March 1807, French, son of a returned émigré.

Goudchaux, Sous-lieutenant, 21 January 1807, French, Jew, volunteer.

Milewski, Sous-lieutenant, 13 July 1807, Pole, coming from his father's lands.

Lipinski, Sous-lieutenant, 27 July 1807, Pole, coming from his father's lands.

Kaiser, Sous-lieutenant, 6 June 1807, Swiss, volunteer.

Migeon, Sous-lieutenant, 1 April 1807, French, returned émigré.

Lux, Sous-lieutenant, 26 April, Pole with prior French service.

Medical Officers

Total

Surgeon Majors:
    Dupont, Kuhl, Filliot and Kurtz   4

Aides-majors:
    Belhomme, Ducasse, Tous-saint, Crétien, Schoy   6
    and Collard.

Sous aides-majors:
    Dietrich, Hamchenski, Heuraut, Guesdon, Harte,
    Rosemberg, Delvaigne, and Hermanns.   8

Recapitulation

Staff Officers	16
Captains	48
Lieutenants	49
Sous-Lieutenants	52
Sub-Total	165
Total	183

In the total number of officers listed above are a number of officers who were named for the 2nd Legion, but who were received into the 1st Legion when they arrived.

Uniforms of the Legion of the North

According to J.R. Elting, *Napoleonic Uniforms*, Vol 2, Plates 71 and 72, by D. Knotel, the Legion of the North was issued new uniforms made from captured materials. The overall uniform was dark blue with red turnbacks, lapels, cuffs, and red piping on the collar. They wore black shoes and black canvas gaiters. The grenadiers, fusiliers, and voltigeurs wore the square topped czapka hat. The grenadiers wore red epaulettes, and red cords and plumes on the czapka. The edges of the czapka top were edged with red. The pompon at the base of the plume had a blue center, surrounded by red, then by white in concentric rings. The czapka had a brass, sunburst plaque over the black visor.

The fusiliers wore blue plume with white trim on the edges of the czapka, and white cords and knots. They wore blue-fringed epaulettes. It is to be presumed that the voltigeurs czapka had a yellow plume, cords and knots, and edging. They wore bright, green-fringed epaulettes.

Elting goes on to say that the officers and soldiers were probably equipped with bicorns prior to the issuance of the czapka.

The uniform Elting describes is probably the uniform that was issued to the Legion after it became part of the army of the Grand Duchy of Warsaw as the czapka was the standard headgear worn by them and it would not have been available to the Legion prior to its entering Polish service.

In contrast, on page 11 Coqueugniot says that they were equipped with shakos. It is most likely that they had some headgear prior to this and that would probably have been a bicorn. The rest of Coqueugniot description lacks any identification of the colors, which makes comparing it to the descriptions provided by Elting and Knotel impossible.

Légion du Nord, Fusilier, 1807.

A fusilier in the Legion of the Nord, 1807 *by Knotel*

Légion du Nord, Grenadier, 1806

Grenadier, Legion du Nord, 1806 *by Knotel*

Extract from L'Armée du Duché de Varsovie, by J. V. Chelminski, pp. 30-31

The investment of the fortress of Danzig began on 14 March, under the direction of General Schramm. At the beginning of the siege, Marshal Lefebvre had, under his orders, around 18,000 men: Dombrowski's division, with 6,000 men including 2,500 men of the Legion of the North, almost all of whom were Polish, 2,200 Baden troops, 5,000 Saxons, and only 3,000 French among whom were 600 sappers from the engineers, an elite troop; the siege artillery only arrived later.

The Polish troops of Dombrowski had ardor and zeal but were unaccustomed to war; the Legion of the North, full of élan, lacked solidity and disbanded on the first shock [of combat]. Marshal Lefebvre also treated his auxiliaries harshly, and he complained in the most outrageous terms. The Emperor addressed him on these subjects with strong remonstrations and counseled him to correct them. "Have," he said, "the indulgence for these young soldiers who are just starting. Tell them that they are good soldiers, which is the best way to make them into good soldiers."

On 7 May the occupation of the island of Holm, carried by force, permitted the completion of the investment of the fortress. The Legion of the North, with the Baden, contributed to the success of this operation; a column of soldiers from these two forces threw themselves boldly against the enemy posts, disarmed them, and captured 200 men and 200 artillery horses. "The Legion of the North, said the 70th Bulletin, and Prince Radziwill who commanded it, distinguished themselves."

On 15 May a vigorous effort by the Russians to break the blockade, was stopped by General Schramm; Captain Sokolnicki, who had already distinguished himself during the siege, and his Polish cavalry, distinguished themselves during this engagement. The siege continued; the corps of Marshals Lannes and Mortier cooperated with the work of the siege forces to hasten a conclusion of the siege. Finally, on the 26th, at the moment when it was decided to launch an assault, the fortress capitulated, and, Marshal Lefebvre made his entry into the fortress. The immense provisions accumulated in the city were taken by the victors; in addition, 4,000 Poles, who served in the Prussian regiments of the garrison, swelled the ranks of the Polish regiments.

In a foot on page 35, one finds:

The 5th Regiment was the old Legion of the North, which had distinguished itself at the siege to Milhaud's division, it remained with it during the seige of Danzig. Attached to it as a garrison for the fortress. It was in the pay of France. After the organization it was given a choice to remain in French service or to enter service of the Duchy. On 11 August, the Legion was brought together in a review, and each company, consulted on its choice, cried, "Vive la Pologne! (Niech z y je Polksa!) [Long live Poland!]" It was only the French officers of the Legion who asked to return to the French Army. The Legion took the name the 5th Regiment of the Duchy and arrived in Warsaw in the beginning of May 1808.

Subsequent history of the 5th Regiment

On 1 January 1809 the regiment, still under Prince Radziwiłł, contained 1,933 men and was stationed in Lissa and Czestochowa. After a year of garrison duty, the 5th Regiment found itself in action again. This time against the Austrians who invaded the newly formed Grand Duchy of Warsaw.

The 5th Regiment first saw action on April 18th Raszyn, where the Polish army fought the Austrian army to a stand still. On the 17th the 3rd Battalion of the 5th Regiment was part of the garrison of Czestochowa when it was taken under siege by the Austrians. On 15 May 3 companies of the regiment participated in a sortie against the Austrians. This sortie and other actions convinced the Austrians to lift the siege during the night of 16/17 May. Apparently, the regiment was not involved in any other significant actions.

On 15 January 1812 the 5th Regiment contained 78 officers and 2,866 non-commissioned officers and men. When Napoleon invaded Russia the regiment was

The 5th Regiment, 1809 by *Jan Chełminski*

assigned to the X Corps, under Marshal McDonald. It was in the 1st Brigade of the 7th division, having 85 officers and 2,553 non-commissioned officers and men on 1 August 1812. It did not participate in any major battles of the 1812 campaign and survived it quite intact, having 80 officers and 1,497 non-commissioned officers still in its ranks in January 1813, while it was serving as part of the garrison of Danzig. On 1 March 1813, still forming part of the Danzig garrison, the regiment had 63 officers and 1731 non-commissioned officers and men. The regiment was part of the garrison when it was besieged by the Allies and remained there the fortress fell to the Allies on 2 January 1814. The 5th Regiment took an active part in the defense of the fortress and participated in at least one sortie. When the city capitulated the Polish members of the garrison were returned to Poland. Their fate after was not ascertained, but it is highly probable that the soldiers were taken into Russian service.

It is ironic how the first major action of the Legion of the North was at Danzig and its last action was there as well. One must wonder if its soldiers sensed that same irony.

Battles and Skirmishes:

1809: around *Czestochowa; Raszyn* 18 April; *Grzybów* 18 May; *Tylża* 24 June; *Żarnowiec* 11 July.

1812: *Dynaburg* 30 July; *Skirmish* 28 September; *Gansenburg* 1 October; *Skirmish* 2, 7, 15 and 23 October; *Neugat* 7 November; *Walkow* 16 November; Skirmishes 17, 18. November, 16 December, 18, 21 December; *Tylża* 28 December; Skirmish 31 December; 1813: Several important skirmishes around Gdańsk: 29 January; 5, 24 March; 15, 27 April; 1 November.

Notable Mentions

Jan Henryk Dąbrowski (1755 - 1818) was born in Poland but grew up in Saxony where his father was an officer in the Saxon Army. He joined the Saxon Horse Guards and fought in the War of Bavarian Succession. He answered the call for Polish Officers to return to Poland following the Four-Year Sejm. He fought in the War of the Constitution in 1792 and the Kosciuszko Uprising. He was part of the radical faction of the army and had an on-going feud with Prince Józef Poniatowski. He promoted the restoration of Poland and formed the Polish Legions in Italy in 1795. He fought in French and Itlaian service until the Treaty of Tilsit, then joined the army of the Duchy of Warsaw. He participated in the Napoleonic Wars, taking part in the Polish-Austrian War and the French invasion of Russia until 1813. After Napoleon's defeat, he accepted a senatorial position in the Russian-backed Congress Poland and was one of the organizers of the Army of Congress Poland.

Antoni Amilkar Kosiński (1769 - 1823) entered the military in 1792. After the Second Partition, he participated in a conspiracy to prepare an uprising in Ukraine but escaped to Tadeusz Kościuszko. During the Kościuszko Uprising he attained the rank of captain in the defense of Warsaw. After the fall of Poland he fought in the French Army, then help supported General Dąbrowski in creating the Polish Legions but resigned his commission over the dispositions of the Legion to San Domingo in 1803. Kosiński helped Dąbrowski organize troops in Poland in 1806, and in 1807 he fought with the Prussians, participated in the offensive against Gdańsk and the capture of Tczew. He replaced the wounded Dąbrowski and led his divisions to Gdańsk, he also took part in the siege of the city. During the campaign in 1809, he was the governor of Warsaw, then the commander of the Polish units in the battles of Szczekociny and Żarnowiec. From 1811, he was a major general. In 1812, he organized defense against the Russians on the Bug line. After the Congress of Vienna, he organized Polish National Guard in the Grand Duchy of Poznań.

Prince Józef Poniatowski (1763 – 1813) was the nephew of the last King of Poland, Stanislaus Augustus Poniatowski. He joined the Austrian Army in 1780 attaining the rank of colonel. He joined the Polish army in 1789 at the request of his uncle, being awarded the rank of major general and commander of the Royal Guards He took part in the Polish–Russian War of 1792, leading the crown forces at the victorious Battle of Zieleńce. After the king's support for the Targowica Confederation of 1792, Poniatowski felt compelled to resign. In 1794 he participated in the Kościuszko Uprising, took charge of defending Warsaw and was exile by Russian authorities until 1798. He developed a rivalry with General Dąbrowski on the conduct of the war and this carried over into the Napoleonic era.

Poniatowski was appointed the minister of war for the Duchy of Warsaw in 1807. He commanded the Duchy's forces during the Austro-Polish War of 1809 and achieved tactical success over a larger and more experienced Austrian force in the Battle of Raszyn. He was a staunch supporter of Napoleon I and died at the battle of Leipzig after receiving a Marshal's baton.

Prince Michał Radziwiłł (1778 – 1850) was a member of the Polish–Lithuanian nobility and a member of the powerful Radziwiłł family. He took part in the Kościuszko Uprising of 1794. In January 1807 he organized the Legion-Du-Nord as colonel and commander and later the 5th regiment of infantry (Duchy of Warsaw). He took part in the siege of Gdańsk and was later stationed with his regiment in Gdańsk. In 1811 he was made general-de-brigade. In the campaign of 1812, he commanded the brigade of infantry in the "Polish division" of general Grandjean in the Jacques MacDonald's 10th Army Corps. He later took part in defense of Gdańsk in 1813 (under general Jean Rapp) and when fortress capitulated, he was taken prisoner. In 1815 he resigned his commission and settled at his estate of Nieborów. In the 1830-31 November Uprising he Commander-in-Chief of the Polish forces for a short time and the Polish commander in the Battle of Grochów. After the failing of the uprising, he was exiled to Yaroslavl in Russia. In 1836 returned to Poland and died in Warsaw.

Józef Zajączek (1752 –1826) entered the Army of the Polish–Lithuanian Commonwealth, as an aide-de-camp to hetman Franciszek Ksawery Branicki. He was initially associated with the conservative elements in the Commonwealth before joining the liberal opposition during the Great Sejm in 1790. He became a radical supporter of the Constitution of 3 May 1791. He rose to the rank of a general, participating in the Polish–Russian War of 1792 and Kościuszko Uprising. After the partitions of Poland, he joined the Napoleonic Army, and was a general in Napoleon's forces until he was wounded and captured during Napoleon's invasion of Russia in 1812. From 1815 he became involved in the governance of the Congress Kingdom of Poland, becoming its first Viceroy. He was constantly engaged in conspiracies against the authorities.

INDEX

Larzaer Coqueugnoit

Look for more books from Winged Hussar Publishing, LLC – E-books, paperbacks and Limited-Edition hardcovers. The best in history, science fiction and fantasy at:

https://www. wingedhussarpublishing.com
https://www.whpsupplyroom.com

or follow us on Facebook at:

Winged Hussar Publishing LLC

Or on twitter at:

WingHusPubLLC
For information and upcoming publications

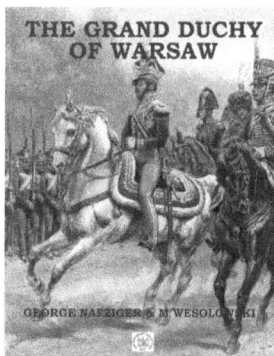  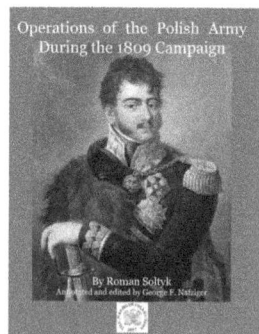

.

www.ingramcontent.com/pod-product-compliance
Lightning Source LLC
Chambersburg PA
CBHW030502100426
42813CB00002B/311